I DIED FOR BEAUTY

I DIED FOR BEAUTY

Dorothy Wrinch and the
Cultures of Science

Marjorie Senechal

OXFORD
UNIVERSITY PRESS

OXFORD
UNIVERSITY PRESS

Oxford University Press is a department of the University of Oxford. It furthers the University's objective of excellence in research, scholarship, and education by publishing worldwide.

Oxford   New York
Auckland   Cape Town   Dar es Salaam   Hong Kong   Karachi
Kuala Lumpur   Madrid   Melbourne   Mexico City   Nairobi
New Delhi   Shanghai   Taipei   Toronto

With offices in
Argentina   Austria   Brazil   Chile   Czech Republic   France   Greece
Guatemala   Hungary   Italy   Japan   Poland   Portugal   Singapore
South Korea   Switzerland   Thailand   Turkey   Ukraine   Vietnam

Oxford is a registered trademark of Oxford University Press in the UK and certain other countries.

Published in the United States of America by
Oxford University Press
198 Madison Avenue, New York, NY 10016

Library of Congress Cataloging-in-Publication Data

Senechal, Marjorie.
I died for beauty : Dorothy Wrinch and the cultures of science / Marjorie Senechal.
   p. cm.
Summary: "In the vein of A Beautiful Mind, The Man Who Loved Only Numbers, and Rosalind Franklin: The Dark Lady of DNA, this volume tells the poignant story of the brilliant, colorful, controversial mathematician named Dorothy Wrinch. Drawing on her own personal and professional relationship with Wrinch and archives in the United States, Canada, and England, Marjorie Senechal explores the life and work of this provocative, scintillating mind. Senechal portrays a woman who was learned, restless, imperious, exacting, critical, witty, and kind. A young disciple of Bertrand Russell while at Cambridge, the first woman to receive a doctor of science degree from Oxford University, Wrinch's contributions to mathematical physics, philosophy, probability theory, genetics, protein structure, and crystallography were anything but inconsequential. But Wrinch, a complicated and ultimately tragic figure, is remembered today for her much-publicized feud with Linus Pauling over the molecular architecture of proteins. Pauling ultimately won that bitter battle. Yet, Senechal reminds us, some of the giants of mid-century science—including Niels Bohr, Irving Langmuir, D'Arcy Thompson, Harold Urey, and David Harker—took Wrinch's side in the feud. What accounts for her vast if now-forgotten influence? What did these renowned thinkers, in such different fields, hope her model might explain? Senechal presents a sympathetic portrait of the life and work of a luminous but tragically flawed character. At the same time, she illuminates the subtler prejudices Wrinch faced as a feisty woman, profound culture clashes between scientific disciplines, ever-changing notions of symmetry and pattern in science, and the puzzling roles of beauty and truth"—Provided by publisher.
Summary: "A biography of Dorothy Wrinch"—Provided by publisher.
Includes bibliographical references and index.
   ISBN 978-0-19-973259-3 (hardback)
   1. Wrinch, Dorothy, 1894-1976.   2. Mathematicians—Argentina—Biography.
3. Biochemists—Argentina—Biography.   4. Women mathematicians—Biography.   5. Women biochemists—Biography.   I. Title.
   QA29.W75S56 2013
   510.92—dc23
   [B]   2012012253

9 8 7 6 5 4 3
Printed in the United States of America
on acid-free paper

For Carolyn Cohen

CONTENTS

I DIED FOR BEAUTY

DOROTHY WRINCH

Some think it a matter of course that chance
Should starve good men and bad advance,
That if their neighbours figured plain,
As though upon a lighted screen,
No single story would they find
Of an unbroken happy mind,
A finish worthy of the start.

—WILLIAM BUTLER YEATS

1 PROLOGUE

Northampton, Massachusetts, 1970. Like most of the rambling houses Smith College rented to its faculty, 35 West Street had at some time been partitioned in two. Dorothy Wrinch lived in the smaller apartment, the one on the second floor. Her front room was crowded with bookshelves, a piano, and an old sofa with lace doilies. A narrow hall led to a bedroom and an eat-in kitchen in the back. She was 76, old enough to be my grandmother, and like my grandmother she lived alone. But my grandmother played cards and baked cookies. I can't imagine Dorothy doing that or crocheting doilies either.

"That nature keeps some of her secrets longer than others," wrote her friend D'Arcy Thompson, "that she tells the secret of the rainbow and hides that of the northern lights, is a lesson I learned as a boy." Banished to the outskirts of the scientific community, Dorothy wrestled the secret of secrets, the secret of life: the inner logic of protein molecules. Now, near the end of her own long life, she sought the answer in the shapes of their crystals. She needed an assistant; I wanted to understand crystal geometry. For the past two years I'd worked with her, unofficially and unpaid, in my scarce spare time, making models and drawing illustrations for the book she was writing. We met several afternoons a week in her office in the science center. Her life was her work, her work her life.

But one beautiful fall morning I drove to her apartment instead. Dorothy was waiting with a bag of sandwiches and hard-boiled eggs. She had suggested this outing a few weeks before, in the middle of a rare reminiscence. How she'd loved long rambles with friends through the verdant English countryside and picnics on the shores of placid lakes! "Let's go to Lake Norwich on Mountain Day," she'd said. Lake Norwich? I'd never heard of it. Nor did I want to go anywhere on Mountain Day. The annual college holiday celebrating the glorious New England autumn was, for me, a day off to play with my small daughters and grade the math homework piled on my desk. But I agreed to a picnic at the lake because she seemed so eager. And because somehow I didn't feel free to say no.

Dorothy was uncharacteristically silent as I drove along the winding country roads that led to the Berkshire mountains. Maybe the splendid scenery left her speechless, or perhaps she was lost in memories. I found the lake: a Mountain Day postcard, red and gold foothills doubled in the mirror of still blue water. Perfectly still, and perfectly quiet: no rowers, no swimmers, no picnickers. Then we saw the sign: No Trespassing. We drove back and ate the eggs and sandwiches on a desolate stretch of road not far from Smith. Dorothy never mentioned lakes or picnics again.

Recently, I came across an autobiography, long out of print, called *The Tamarisk Tree*.[1]

"During the summer of 1916 Dot Wrinch rang me at Sutton to ask me to go for a weekend walking tour with her, Bertrand Russell and Jean Nicod, one of his best pupils," the author, then Dora Black, recalled inaccurately (the year in fact was 1917). The girls, both exuberant, irreverent, redheads, had finished their studies at Girton College, Cambridge. Now Dot was studying with Russell in London. His pacifist activism had cost him his Cambridge lectureship.

"Russell, as I later learned, took great pleasure in long walks," Dora continued. "The idea was to walk over the downs near Guildford, spend the night at Shere and go back by train to London on Sunday evening. Dot joined the three of us in the train at Surbiton, carrying a large basket of very fine strawberries from her garden. Russell took charge of our route, indicating short cuts to Shere, while Dot kept whispering to me that she was quite certain 'our dear Bert' was entirely wrong and we were in fact heading for Gomshall."

Rabbits bounced among the wildflowers in the North Downs meadows. At 45, Russell looked rather like the Mad Hatter, Dora noted as the quartet rambled along unmarked paths. Her three companions were mathematicians and philosophers; they talked a lot of shop. Nicod "had a type of whimsical humour that delighted me," Russell would write in his own autobiography. "Once I was saying to him that people who learned philosophy should be trying to understand the world, and not only, as in universities, the systems of previous philosophers. 'Yes,' he replied, 'but the systems are so much more interesting than the world.'"

After dinner they chatted in the candlelit parlor. Dora says the inn was in Gomshall; Russell says Shere. But neither forgot what was said. Russell asked his friends, all half his age, what they most desired in life. In Dora's words: "Nicod, who felt himself plagued by a very sceptical disposition, said he thought that he would like to be really absorbed and caught up in some great belief or cause; Dot, I think, also hoped that she would find something entirely absorbing to which to devote her energies. Conscious that my choice was a bit banal, but speaking with sincerity, I said that I supposed I really wanted to marry and have children."

A candlelit evening 95 years ago now. The young people's wishes would be granted—but, like characters in the short story "The Monkey's Paw," with consequences they could not have foreseen and would not have wished. In 1921, Dora Black would become the second Mrs. Russell and, in short order, the mother of four children, but the marriage ended bitterly 12 years later. Jean Nicod "managed to maintain his pacifist objection and survived the war, but only at the great cost of making himself ill and unfit, since in France there was no provision for conscientious objection." He died of tuberculosis at the age of 31.

As for Dot Wrinch, this book is her story.

2 CULTURE CLASH AT COLD SPRING HARBOR

A few years after our aborted picnic, Dorothy left Northampton to live year-round in her summer home in Woods Hole. I visited her from time to time to talk about the book, but she gradually lost interest in it, and anyway, my viewpoint was diverging from hers. In 1974, she suffered a stroke. In December 1975, her only child, Pamela, died in a fire. Dorothy never spoke again. She died in February. Snow crystals muffled the clergyman's graveside eulogy. She would have liked that.

Dorothy left her papers—her notebooks, correspondence of 40 years, reprints, and models—to Smith College. Pamela's husband asked me to organize them for the archivist. And so I slipped into the maelstrom of her life and career, about which she'd told me very little, and into the controversy that still lingers three decades after her death.

I stir the pieces of her story around and around in the dense soup of ambient issues, coaxing them to crystallize, to arrange themselves in a shape I recognize and understand.

The arrangement of atoms in a crystal is the mark of its species. Stack carbon atoms one way, and you get graphite, the soft "lead" in pencils; stack the same atoms another way, and you get diamond, tops on the hardness scale. Table salt is a three-dimensional checkerboard of sodium and chlorine. Silicon and oxygen link in helices to make quartz.

A crystal's outer shape—its faceted geometric form—refracts the nature/nurture conundrum. As the crystal grows, atom joining atom in its characteristic pattern, the "mother liquor" (as chemists call it) nudges the crystal toward the shape that we see. Her action is subtle: trace ingredients in the liquid or its rate of cooling can make some facets grow larger and others disappear. Left to its own devices, a salt crystal becomes a cube; in 2005 the United States government granted patent #20050136131 for a method of growing salt in a shape less likely to clog up the shakers. Quartz grows in prisms, long ones, pointed at one end. But in Herkimer County, New York, the prisms are short and both ends are pointed. *Der Diamant*, a 1911, still-definitive, two-volume treatise on diamonds, has over 200 drawings of uncut diamond crystals with subtly and not-so-subtly different shapes.

Dorothy Wrinch's story crystallizes slowly and its form is unstable. It varies with the teller's temperature and trace elements in the teller's eye. I'm looking for patterns and facets. Where should I begin? "Science too," George Eliot reminds us in *Daniel Deronda*, "reckons backwards as well as forwards, divides his unit into billions, and with his clock-finger at Nought really sets off *in medias res.*"

In medias res, then. In the middle of things. In the eye of the storm.

July 1938, Cold Spring Harbor, Long Island. The Biological Laboratory in this pleasant village is already renowned in scientific circles, though Barbara McClintock hasn't planted corn here yet, and James Watson is only 10 years old. It's playing midwife to a scientific revolution—a revolution as momentous as the Copernican and Darwinian. Inspired by the "molecular vision of life," the Rockefeller Foundation and the lab are jump-starting a *new* biology, crosscutting and interdisciplinary. Warren Weaver, the Foundation's visionary director for Natural Sciences, has just come up with a name for it: *molecular biology*.[1] Scientists jump and crosscut for invitations to Cold Spring Harbor conferences. To arrive here is to have arrived.

Once you've arrived, you can relax a little. Dress is casual, sport is sporting. The tennis court is down the road. The redhead slamming balls across the net is Dorothy Wrinch. She plays athletically and energetically—the same way, friends note, that she plays the piano.

Dorothy is 44, but she looks younger. She signs her letters like this:

$$\left\langle\!\!\left\langle\; \delta\; \right\rangle\!\!\right\rangle$$

and her friends call her Delta.

What is she doing here, in Cold Spring Harbor? Place the emphasis wherever you wish: what is *she* doing here? what is she *doing* here? what is she doing *here*? Someone is asking it that way.

She's *here* because Eric Ponder, director of the Biological Laboratory, has invited her to lecture in a month-long conference, the Sixth Symposium on Quantitative Biology. *Quantitative*: the very name suggests a break with biology's descriptive, compartmentalized past. Each year's symposium treats a

Dorothy Wrinch at Cold Spring Harbor, 1940.   Courtesy Cold Spring Harbor Laboratory Archives.

cutting-edge topic. In 1938, proteins are as hot as July. Proteins are a topic whose time has come. It's now an established fact: proteins are *molecules*. Dorothy Wrinch is an international superstar of modern protein science. Her model for the molecular architecture of proteins is catalyzing research on both sides of the Atlantic.

Dorothy Wrinch is one of 72 promising, mostly young, scientists who have traveled here by boat, car, and train from England, France, and many states. She's the only mathematician among these chemists, botanists, bacteriologists, physiologists, zoologists, physicists, physicians, and x-ray crystallographers. Fourteen of them work in the Biological Laboratory or in the Rockefeller Institutes for Medical Research in New York or Princeton (and several have in the past); the research of others, including Dorothy's, is supported by Rockefeller Foundation grants.

The conference is informal: just one lecture a day, leaving hours unstructured for casual discussion and sporadic debate. In the lecture hall, at the breakfast table, lunch table, dinner table, swimming in the Harbor, sailing on the Sound, lolling on the lawn, strolling through the woods, even on the tennis court, the scientists talk about proteins. Most of them stay the whole month.

Whatever you think of Dorothy Wrinch—opinions differ—you have to grant she's prodigious. She holds a D.Sc. in mathematics from Oxford—the first Doctor of Science degree Oxford University ever gave to a woman. She also holds master's degrees from Cambridge University and the University of London, and a D.Sc. from the latter. Dorothy has been teaching mathematics at Oxford for 14 years. She lists 50 papers in mathematics, philosophy, and scientific method on her resume, all published in sterling journals. She doesn't list her book *The Retreat from Parenthood*; she wrote it under a pseudonym.

But Dorothy is no grind. She loves company; she's vivacious. Her wit is mordant. She has something to say on almost every subject: Shakespeare, *Don Quixote*, the *Kama Sutra*, Schoenberg's twelve tones, Roger Fry's Omega Workshop, Russell's philosophy, quantum mechanics and aerodynamics, the continuum hypothesis, the need for national child-rearing services, and the looming war. She can say it in Oxford English, Parisian French, Viennese German.

Dorothy doesn't talk about her husband John, his madness and institutionalization, their recent divorce. She doesn't talk about her struggle to support their daughter, now 11 years old. She keeps her loneliness to herself. She shows no one the anguished jottings, the painfully honest lists of pros and cons she makes for every big decision.

Sixty years later, in Stockholm for the Nobel ceremony honoring his discoveries (they led to Viagra), Robert Furchgott, the youngest participant in the Cold Spring Harbor gathering of 1938, would remember that summer as his metamorphosis from student to professional. His professor, an expert on the protein egg albumen, had brought him along; in exchange for room and board he tended the lantern slide projector in the lecture hall. The symposium, packed with famous scientists, was a heady experience for a 22-year-old aspiring chemist.

Irving Langmuir was the most famous chemist at the gathering, maybe the most famous physical chemist in the world. He'd already invented the gas-filled incandescent lamp and discovered atomic hydrogen. He'd won the Nichols Medal (twice), the Hughes Medal, the Rumford Medal, the Cannizzaro Prize, the Perkin Medal, the School of Mines Medal, the Chandler Medal, the Willard Gibbs Medal, the Popular Science Monthly

Award, the Franklin Medal, the Holly Medal, and the John Scott Award. And of course, the Nobel. No one else there had won a Nobel yet.

Tall, loquacious, larger than life, Langmuir climbed mountains and flew his own plane. And when a trembling neophyte dared to challenge his remarks in the lecture hall, in front of everyone, the great man replied thoughtfully, as he would to a colleague. Bob Furchgott, the neophyte, never forgot that.

Langmuir was Wrinch's staunchest supporter. They'd written several papers together and would write several more. The next year, he nominated her for a Nobel Prize.

He tended the lantern slide projector…how quaint, how ancient, that sounds, in our digital era, our world of streaming video and jpegs, YouTube and PowerPoint.

Lantern slides—glass slides, 3.5 x 4 inches, with photographic images transferred to them by any of several methods—were patented in 1850. The invention brought the inventors, William and Frederick Langenheim of Philadelphia, a medal at the first of the great world fairs, the Great Exhibition at the Crystal Palace in London in 1851. The brothers meant to entertain, nothing more. But the impact of these replicable, portable slides was far greater: the lantern slide brought the world to the lecture hall. In its century-long heyday, from the invention of photography to the Second World War, the "magic lantern" transformed the transmission of art and science.

The old boxy lantern slide projectors are museum pieces now, and lantern slides are collectors' items. My Google search for "lantern slides" turned up 18,200 images in .18 seconds: grape harvesters in Germany, circa 1900; villagers in India; a tiger in a London zoo; white-hatted archaeologists digging in a desert; and faded images of the great pyramids. No mathematics, chemistry, botany, bacteriology, physiology, zoology, physics, x-ray crystallography. Not the stuff of lectures on proteins.

Some years ago, the director of the Smith College Science Center implored us, the faculty, to clear out our sub-basement storage cages. I poked around the ill-lit jumble for an hour or so, retrieving a few long-forgotten models, tossing old notes in the trash. I found lantern slides in a green metal cabinet; they weren't mine to throw away. I gathered up my models and relocked the cage.

Now, Cold Spring Harbor on my mind, I remember that green metal cabinet. Is it still there? Are the slides still inside? Might some of them be Dorothy's?

The storage cage is still stuffed with boxes, broken furniture, unrecognizable pieces of forgotten instruments. I push my way around chairs with missing legs, decrepit desks, outmoded balance scales, dials pointing to nowhere, abandoned machine parts. The green metal cabinet still rests against the wall. And yes, it still holds the slides in its top two drawers—1374 glass slides, each in a yellowing envelope, each carefully numbered by hand. A mosquito, a grasshopper, a corn earworm, trees, vegetation, landscapes. The ice sheet over North America, a map of the location of dinosaur remains. Workers transplanting trees, an onion field, a sisal hemp plantation in Uganda, portraits of Swammerdam and Darwin and other bearded venerables, and gastric pouches patented December 29, 1896. No crystals, no molecules. These are the staples of Smith science courses long ago, the biology the scientists at Cold Spring Harbor learned in school.

As I grope my way back through the cluttered cage, I spot a cardboard box on a high shelf of metal staging. It's very heavy; I can scarcely lift it down. It's filled with lantern slides. These slides have no numbers, and most have no envelope. I browse through them: models, crystals, diffraction patterns. The images are elegant, concise, precise. My heart skips a beat; then tears blur my eyes: these are Dorothy's slides. She must have stashed them here when the science center opened, to great fanfare, in 1965. Her new office was small and by then lantern slides were history, supplanted by new technology: Kodak carousels, overhead projectors. She would never use her lantern slides again.

The oldest slides in the box are hand-made: disintegrating negatives clamped between glass plates, bound with red or black tape. I hold one up to the light. The glass is cracked, the aged tape disintegrating.

Dorothy's protein model. Simple, beautiful, elegant. The geometrical objet d'art that catalyzed research on both sides of the Atlantic.

Today all freshmen in Bio 101 know that a protein is "analogous to a string of Christmas tree lights, with the wire corresponding to the repetitive (polymer) main chain, and the sequence of colors of the lights to the individuality of the sequence of (amino acid) side chains,"[2] just as DNA is a double helix, and genomes Я us.

The case is long since closed. "Proteins are now so well understood and so much a part of our knowledge that it is almost impossible to put ourselves in the position of the participants in the 1938 Symposium," says the Cold Spring Harbor website.

Almost impossible to put ourselves in the position… but let's try. In 1938, the case is open. We are aware of the chain hypothesis, of course, but there are other, maybe better hypotheses, like Dorothy's. Experiment will decide

Dorothy Wrinch's cracked lantern slide showing her protein model. The model was constructed and photographed in Niels Bohr's laboratory in Copenhagen; the lantern slide was made for Irving Langmuir, 1940.

the matter someday, but so far it has had little to say. Meanwhile, we debate the merits and demerits of competing suggestions.

A valid hypothesis must fit the facts. One fact is: proteins have specific jobs to do. Take insulin, for example (it's the example everyone takes). Insulin has given diabetics all over the world a new lease on life. "Who could have imagined that an assortment of amino acids put together in a certain combination could exert such a profound physiological effect?" muses Vincent du Vigneaud. He heads Cornell University's biochemistry department; the Nobel Prize in Chemistry will come his way in 1955.

A "certain combination"—du Vigneaud knows that's the nub of it. Whatever structure you propose for the insulin molecule must be able to account for that profound physiological effect. If you say insulin is a chain, a string of 10,000 Christmas tree lights, then tell us precisely how it works. Meanwhile, we will keep an open mind. Maybe insulin isn't a chain at all.

Experiment has settled the case of fibrous proteins like silk, wool, linen, mammalian hair, hoof, feather, and horn. Fibrous proteins are solid state; they're built of submicroscopic crystallites. Crystallites diffract x-rays. Bill

Astbury's diffraction photographs show that these proteins are chainlike. No one disputes that.

But insulin isn't a fibrous protein, it's "globular." In its "native" or active state, its overall shape is more or less round. Hemoglobin, the albumins, pepsin, gamma globulin, and other proteins critical for life are globular too. Their structure is the prize. Are globular proteins chains wound up like golf balls? Or is their architecture something else entirely? That's the big question here at Cold Spring Harbor.

X-ray diffraction hasn't helped, so far. Globular proteins can be crystallized, but that takes care and skill and luck—it's something of an art form. And these crystals are delicate, their diffraction patterns bewildering, in 1938.

If x-rays don't reveal the molecular architecture of globular proteins, cooking might. Anyone who's fried an egg or beaten an egg white knows that egg albumen's native state is transformed by heating or beating. The protein unfolds and flattens out. The unfolded—"denatured"—state can be studied by chemical techniques. Maybe the protein's native state can be inferred from its denatured remains.

Bill Astbury, the British physicist who x-rayed the fibrous proteins, thinks all proteins are one big happy family. Denatured globular proteins look fiberlike to me, he says.

Alfred Mirsky, a protein chemist at the Rockefeller Institute in New York, disagrees. "To regard fibre formation as the criterion of denaturation and to 'reserve' the term 'denaturation' for the fibrous state is apt to be misleading."

Others doubt there is such a thing as *the* denatured state.

"It is to be expected that a denatured protein is in general not a definite substance but is rather a mixture of many," says Irving Langmuir.

Henry Bull from Northwestern, Bob Furchgott's professor, agrees. "The term denaturation has no meaning except in connection with operations which one performs."

Dorothy Wrinch's model accounts for different denatured states. "The fact that surface pattern can fragment into line patterns implies that line polypeptides, as well as pieces of fabric, can result from the process of denaturation," she says.

What is *she* doing here?

"Sir," said Samuel Johnson 200 years before, "a woman's preaching is like a dog's walking on his hinder legs. It is not done well; but you are surprised to find it done at all." Dorothy is not the only woman at Cold Spring

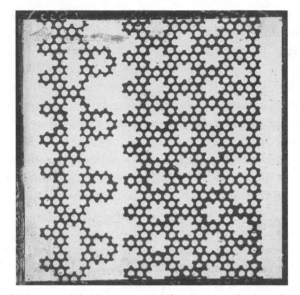

Dorothy Wrinch's lantern slide showing cyclol fragments as denatured proteins.

Harbor. The symposium group photograph shows smiling wives in summery dresses and children large and small. (Eleven-year-old Pamela Wrinch is not among them; she's home in England with her "godless godmother," Margery Fry.) But only five of the 72 invited scientists are female, and of these only two, Dorothy and Eloise Jameson, a chemist from Stanford University, will give lectures. Just two women speakers, but it is 1938. You are surprised to find any at all.

Dorothy does it very well. She's a pro. She sizes up her audience, pitches her talk to just the right level. She doesn't read a written text; scientists rarely do. But Dorothy doesn't just not read: she shows. She shows elegant models and exquisitely drawn pictures. Their beauty speaks for her, or rather with her, for her words are eloquent too.

Dorothy dazzles everyone, or almost everyone—the men at her lecture, their wives in the dining room—with her scintillating conversation. But a little bit of Dorothy Wrinch can go a long way. She can get on your nerves. "A queer fish," Warren Weaver of the Rockefeller Foundation records in his diary, "with a kaleidoscopic pattern of ideas, ever shifting and somewhat dizzying."[3] Max Bergmann and Carl Niemann, two chemists here, can't abide her. The feeling is mutual.

But most of those who ask what *she's* doing here have stayed away. Linus Pauling stayed in California. John Desmond Bernal and Dorothy Crowfoot Hodgkin stayed in England. In 1938, we have no e-mail and the snail is slow. The phone down the hall is a party line. We can't dial long distance directly, we have to call the operator, and we do that only in emergencies. It's not exactly calm here in the eye of the storm, but it could be—and will be—worse.

Then what is Dorothy *doing* here? What makes her a superstar of protein science? Why is her model such a sensation?

Step once more in their sandals. It's miserably hot in the lecture hall, though the whirring electric fans do their best. Forget today's jargon: CT, NMR, MRI, and DNA mean nothing at all in 1938; the word "acronym" won't be coined for another five years. We glimpse no connection between heredity and nucleic acids, though Astbury, whose lecture here will long be quoted, insists they're worth looking at. If there really is such a thing as a gene, and some still doubt it, it's probably protein. Just as the hormone insulin has turned out to be. The tomato bushy stunt virus is pure protein too— Wendell Stanley, who is here, created quite a stir when he proved that last year. His picture was in all the papers; he was compared to Louis Pasteur.

Data pour in from Uppsala, Sweden, where Theodor Svedberg is studying proteins with an ultracentrifuge of his own design. "When a slurry, an emulsion, is put into a rapidly rotating motion, its heavier constituents are thrown outwards in the direction of the periphery of the motion," the Nobel Committee explained on awarding him its prize in 1926. "This happens in the most used of all centrifuges, the milk separator, where the skimmed milk is pressed outwards whilst the lighter fat particles, the cream, accumulate inwards and can therefore be separated." With the help of his ultraprecise ultracentrifuge, Svedberg calculates the molecular weights of the separated material. He's studied a great many proteins by now, but he has not found a great many weights. It seems proteins fall into a small number of weight classes. Why is that?

That's not all. Poring over Svedberg's data, Bergmann and Niemann have found unexpected regularities. Invariably, the estimated number of amino acid residues in every protein is a product of powers of 2 and 3! For example, the number for hemoglobin is 576, and $576 = 2^6 \times 3^2$. The pattern goes deeper. The fraction of each *type* of amino acid residue in any protein is also a product of 2s and 3s. Why is that?

The answers are elusive, but we're closing in on them. It's 1938, morning in protein science! We don't know that the dawn is false.

In her student days, in Cambridge, Dorothy and her best friend Dora Black cycled to meetings of the Heretics Society. Twenty years later, she's a heretic still.

Globular proteins aren't chains, says Dorothy, they are *fabrics*. She dubs them "cyclol" fabrics because they're made up of rings. Rings, not chains, of Christmas tree lights. Twinkling hexagons joined together like lace. The fabric folds up into cages like origami. And what folds up, unfolds out. "The disorganization of the compact and orderly structure of the native protein is simply the ripping open and fragmentation of cage structures in general," she says.

Dorothy's cage is symmetrical, nearly Platonic. It explains many facts. It's the first specific model for proteins that anyone, anywhere, has ever devised. I hear the audience gasp.

Is there a protein fabric? That's the title of her lecture, but it's a rhetorical question; Dorothy's sure she's right or at least on the right track. She's quick on her feet, she holds her own in the after-lecture give-and-take of probing comments and questions. She passes her model around the audience; they turn it this way and that. It's a model of her model, actually: a material, elegant, realization of her abstract, elegant idea.

"I am not sure just where the side chains go in your picture," says Lawrence Moyer, a botanist from the University of Minnesota. "Do they point out off the surface of the fabric or do they go into the center?"

"If we take any kind of a fabric structure like the cyclol fabric there should be two kinds of proteins," Dorothy replies. "Those in which the side chains start by emerging from the cage and those which start by penetrating the cage."

"That might explain why the amino groups are not essential for the physiological activity of insulin," says Abraham White, Yale School of Medicine.

Chains, or cages? Fibers, or fragments? We, the assembled scientists, disagree on this and almost everything else. But, happily, we are unanimous on one point: the Bergmann-Niemann formula is of utmost importance. Everyone at Cold Spring Harbor cites it as established fact; everyone is convinced it says something profound about proteins. But what that is, no one knows. No one but Dorothy Wrinch.

"You have various ratios which are multiples of 2 and 3," Conrad Waddington, a geneticist from Edinburgh, says to Niemann. "You do not have ratios of numbers like 5 and 7. Why are some numbers excluded?"

Like Newton, Niemann declines to hypothesize. "I cannot explain it; it is an experimental fact."

Dorothy Wrinch does hypothesize and she *can* explain it. Those 2s and 3s come from the symmetries of the protein fabric: two hexagons share each edge, three share each corner. That's why her exquisite cages come in different sizes, but not just *any* size. When you work out the details, as she has, you find that C1, the smallest cage, accommodates 72 amino acid residues, C2 can hold 288, C3 has 648, and so on; the general formula is $72\,n^2$. Since 72 itself is a product of twos and threes, $2^3 \times 3^2$, the formula $72\,n^2$ will be a power of 2 and 3 when n is 1 or 2 or 3 or 4 or 6 or 8 or 9. That's what the Bergmann-Niemann formula means, says Dorothy. It means that proteins are cages.

"The beauty of mathematics faces you"—I quote the mathematician David Ruelle—"in those moments when the underlying simplicity of a question appears and its meaningless complications can be forgotten. In those moments…some of the meaning hidden in the nature of things is finally revealed."[4]

But the mathematician's "meaningless complications" may be the experimentalist's devil in the details.

The molecular vision of life needs time to focus, longer than a get-together, even one lasting a month. The Cold Spring Harbor Symposium is a three-way culture clash.

Ralph Wyckoff, an x-ray crystallographer from the Lederle Laboratories, pleads for cooperation. Macromolecules are "perhaps the last great blind spot in our knowledge of the material composition of our immediate environment," he says. And biologists and chemists view this blind spot through opposite ends of the telescope. "The biologist enters this region from above through the investigation of microorganisms that become progressively smaller. The physical chemist approaches it from below by finding molecules or by producing aggregates of particles that are bigger and bigger."

"To progress in this field," says Wyckoff, "we must learn to effect a compromise between two disciplines which in a real sense are out of touch with one another."

Not two out-of-touch fields, but three. Dorothy Wrinch, the mathematician, approaches this blind spot from outside the telescope, outside the bounds of space-time. Her fabrics and cages come from Plato's world of abstract geometric forms.

Fabrics, or chains: let's decide that principle first, says Dorothy. Whether a particular protein is this fabric or that chain, my fabric or your chain, we can leave for later. The details—how and where the amino acids pack in, what holds them together—will sort themselves out. *May I ask for criticism specifically directed toward the idea that the essential entity in proteins is a fabric, leaving to one side the question of the nature this fabric must have if it exists?*

No, you may not, they say in effect. But in fact, they don't even hear her.

Three approaches to one blind spot, three long scientific traditions. In the seventeenth century, Robert Hooke in London looked through his 20x microscope and drew what he saw with "a sincere hand and faithful eye." The edge of a razor, tiny crystals, a flea, watered silk, a sliver of linen, a thin slice of cork: his monumental *Micrographia* was an instant best seller. Jonathan Swift dashed off a witty ditty:

> So nat'ralists observe, a flea
> Hath smaller fleas that on him prey,
> And these have smaller fleas that bite 'em,
> And so proceed ad infinitum.

Hooke coined the word "cell" in his description of cork:

> I could exceedingly plainly perceive it to be all perforated and porous, much like a Honey-comb...these pores, or cells...were indeed the first *microscopical* pores I ever saw, and perhaps, that were ever seen.

Now, in 1938, vastly more powerful microscopes show much the same thing. In Cold Spring Harbor Wanda Farr, a botanist from the Boyce Thompson Institute, sees fabrics everywhere: "In our microscopic study of plant cell membranes we are constantly confronted with their fabric-like appearance...The fabric-like nature of these first visible aggregations may be significant as an indication of one of nature's observable methods in building up organic structures."

The blind spot looks blank from the telescope's other end. The words "atom" and "molecule," used interchangeably for millennia, acquired their present meanings only in 1871, the year of the Great Chicago Fire. That the atoms composing a molecule are arranged in a specific way was another hard sell. As late as 1874, J. H. van't Hoff (who, in 1901, would be the first Nobel laureate in chemistry) was ridiculed for suggesting that the bonds of the carbon atom point toward the corners of a tetrahedron. The atomic scale was still hazy in 1900: "atom" meant one thing to chemists, another to physicists, and yet another to crystallographers, while others doubted the reality of atoms at all. Einstein's explanation of Brownian motion in 1905 helped end the doubts and confusion. But how nature puts atoms together— "producing aggregates of particles that are bigger and bigger"—remains, in 1938, unimaginable.

Dorothy doesn't look through either end of the telescope: she doesn't need it at all. Galileo, whose own telescope opened the grand book of the universe for all to see, taught her to read the book without it. This grand book is written in the language of mathematics, he said. "Its characters are triangles, circles, and other geometric figures, without which we cannot understand a word of it, but only wander aimlessly in a dark labyrinth." Geometry is the language of the universe and everything in it. Including proteins, says Dorothy.

Two decades before C. P. Snow made "the two cultures" a household phrase, he observed the gulf between two scientific cultures, chemistry and physics. Snow knew whereof he wrote: he'd begun his career as a physicist at Cambridge; he knew Bernal and Dorothy and all the rest of their crowd. His 1934 novel, *The Search*, is a *roman á plusiers clefs*. Arthur Miles, Snow's stand-in narrator, is an ambitious young physicist who dreams of, and schemes for, a research institute of his own. When an influential physics colleague lectures to the Chemical Society, a "long and petulant discussion" ensues. As the two men leave the meeting, the politically savvy Miles tries to butter him up.

"A queer, fierce, quarrelsome crowd they are," I said. "Why is chemistry the most conservative of sciences?"
"Because it's got no mathematical basis," he said promptly...
"You mean," I said, "that there's nothing to test the new ideas by? And the old ones have all the force of tradition behind them."

Cross-disciplinary discussions at Cold Spring Harbor are long and petulant too.

"It seems to me that Wrinch's theories have great interest for us and that it is to be hoped that conclusive evidence may be found," Edwin Cohn, Harvard School of Medicine, remarks politely, "but I do not think that there is any evidence for the spatial arrangement."

"We should stay on experimental ground, and if someone claims that there are other linkages in proteins, he should try to find an experimental way to prove the presence of these linkages," Bergmann says acidly.

"I wished to refer to the much broader issue as to whether or not it is reasonable to postulate that some fabric or other is the essential entity in protein structures," Dorothy insists yet again.

Up and down the hierarchical levels, back and forth between main chain and side chain: the arguments rage for the month.

"There is, it would seem, in the dimensional scale of the world a kind of delicate meeting place between imagination and knowledge, a point arrived at by diminishing large things and enlarging small ones, that is intrinsically artistic," Nabokov will write.

If she's right, Dorothy's elegant protein fabric is that hoped-for meeting ground of imagination and knowledge: lace for the biologists, twinkling lights for the chemists, and Plato gets the cage. Her model accounts for many of the myriad questions that swirl around proteins. It folds and unfolds. It gives meaning to the Bergmann-Niemann formula. It accounts for Svedberg's data on weights. But Dorothy can no more explain why a protein fabric, if there is one, *should* fold and unfold as she says it must, than her doubters can explain how a chain could do better.

The statistician George Box will later quip, "All models are wrong, but some are useful." But by then the role of models will be better understood. In 1938, the mathematician and the chemists talk past one another. It is not so very long after *The Search*.

On and on they argue—sometimes head-on. And at the end of the month they all go home, all but young Furchgott, who stays on to work in the lab.

The published volume of symposium lectures and discussions, from which I've drawn most of this account, "provides an illuminating snapshot of the uncertainties and confusion that surrounded the nature of proteins." The false dawn faded to darkness as many of the Roman candles that lit these lively debates fizzled and landed in history's dustbin.

Svedberg was wrong: proteins do not belong to a small number of weight classes.

Stanley was wrong: viruses are not pure proteins.

Bergmann and Niemann were wrong: their formula does not hold for most proteins.

And Dorothy was wrong: globular proteins are not folded fabrics.

Yet—and yet. Theodor Svedberg is still revered for his earlier discoveries. Wendell Stanley won a Nobel Prize in 1946 for his work on the virus.[5] Max Bergmann remained the Rockefeller Institute's protein expert until his death in 1944. Carl Niemann teamed up with Linus Pauling.

Dorothy Wrinch made a difference in half a dozen fields: mathematics, philosophy, seismology, genetics, protein structure, and the theory of x-ray diffraction.

So why was she marginalized? Why is she forgotten?

I stir the pieces of her story around and around, again and again.

3   SYMMETRY FESTIVAL

Autumn 1969. I have to do it sometime; it might as well be now. One day each week from September to May, vacations excepted, the scientists teaching at this small New England college descend on the faculty club by Paradise Pond to munch while one of them lectures. Any subject I like, the lecture chairman reassures me, as long as it's of interest to everyone.

Everyone in this blur of graying physicists, querulous chemists, geologists in boots, smiling psychologists, astronomers with accents, microbiologists and botanists and zoologists still grumbling over their departments' recent merger? All of them, and my fellow mathematicians too? I'm new here; I scarcely know anyone.

Everyone.

Not my Ph.D. dissertation, then. It interests no one, not even me: I've abandoned that topic, with no loss to the field. And not my new field, geometrical crystallography: I don't know enough yet, and this is not the time to make a fool of myself. Maybe the new course on symmetry I'm planning for the spring—it passes the breadth test, no mean feat for a math course. My students will grow crystals, find symmetries in their atomic patterns, see the effects through polarized light. We'll study symmetry in animals and plants, and in the solar system. We'll track symmetry in music, art, literature, dance. Symmetry is a vine, its shoots spread everywhere. Mathematics, Queen of the Sciences, is the node to which we'll trace them back. Yes, my course fits the bill; it has something for everyone except the psychologists. Years later I will understand that it was only for the psychologists.

I have to send in a title. In the faculty coffee lounge, I agonize aloud. "Plans for a Course on Symmetry"—boring, boring. "A Course on Symmetry"—nearly as bad. "Symmetry"—appalling, grandiose. A wise old chemist looks up from his newspaper. "Call it anything," Professor Soffer says. "Just add 'And You.' They'll come."

I don't, but they come anyway. The new science center is just across the road from the club. The scientists have been scattered across the campus for 110 years, sharing corridors and even offices with historians, classicists, artists, language teachers, philosophers. But times are changing in academia, as in the streets. The tectonic plates called disciplines are inching faster now, colliding, crumbling, remerging. These times call for cross-pollination, cooperation, sharing. The new building complex is designed to bring this about. Glassed-in bridges, narrow corridors, one library, one coffee lounge, one bank of mail slots, one secretarial pool with one mimeograph machine. And, soon, one room-sized computer.

One community. At lunchtime, the community is hungry.

I'm terrified. I've given talks before, but only to mathematicians. The elders in this eclectic crowd hold the keys to my future: reappointment, tenure, promotion. What will they make of my sprawling, experimental course plan? It suits the changing times, it's philosophically correct. But, as Karl Marx said à propos of something else, "New superior relations of production never replace older ones before the material conditions for their existence have matured." Maturing takes more than moving in. Disciplinary boundaries are still heavily mined. And I propose to traipse blithely across them.

The day comes, as days do. Despite severe stage fright, I get through the talk. I say more or less what I more or less planned to say. I put my acetate slides on the overhead projector right side up and in the right order. I don't drop my red, purple, brown, green, blue, and black marking pens. No one snores, no one stares out the windows at the pond…And then someone asks me a question. What it is, I'll never know. Blindsided by nausea, I hear nothing but the pounding in my woozy head. I grab the back of a chair and hold on. The audience, oblivious to my agony, gazes sphinxlike as I sway, mute, mourning my stillborn career. Then a familiar British voice chirps from across the room, "A very interesting question! May I say a few words?" All swivel. Dorothy Wrinch takes the floor and holds it until I find my voice and raise my head. "This will be a marvelous course," she says as she sits down. If some think otherwise, they don't dare say so out loud.

I didn't meet Dorothy in my first year at Smith; I didn't meet anyone. It was not the optimal moment to start an academic career. I had a two-year-old daughter and a baby on the way when a replacement position materialized suddenly, and I had to *carpe* the *diem*. In those last years before a storm of student protests yanked off Smith's white gloves, its antiquated, idiosyncratic schedule of classes remained in force: some met Monday-Tuesday-Wednesday, others Thursday-Friday-Saturday. Debutantes with heavy weekends in Boston or New York were grateful for the three-day option; professors with the requisite clout were happy too. But I was the lowest totem on the math department pole. Six days a week, Monday through Saturday: I could take it or leave it. Monday through Friday, I dropped Diana off for morning daycare (my husband handled Saturdays), drove the nine miles from my home in Amherst over the Coolidge Bridge to Northampton and up the hill to Smith. I taught my classes, saw students at office hours, and returned home as quickly as I could. Jenna, considerate then as now, arrived the first day of spring vacation. My department chair brought the seniors' honors theses to my house a few days later, to save me the trip to the college.

The time pressures eased a bit the next year. A marvelous Mrs. Peterson came every day to care for the girls. I taught my four courses three days in a row and held office hours *et cetera, et cetera,* in the other two. I had time for a lunch break, time to check the mailbox, time to browse in the library; I could even stop by the lounge for coffee once in a while. But still I didn't meet Dorothy. Not in the lounge, not in the stacks, not in the corridors, not at the mail station. If not for a rare book in the art library and an inquisitive professor, I might never have met her at all.

I found that rare book, and Dorothy Wrinch, at the end of a long chain of footnotes. My quest began a few years before, when a Brazilian friend gave me a ruby, my birthstone. The ruby was large and dark red, with glittering facets. I was surprised by such a generous gift, but maybe it wasn't *that* extravagant. Brazil is the land of magnificent crystals. It even has a state named Minas Gerais, or General Mines. I brought my ruby to a jeweler in Northampton to set in a ring. Mr. Gare took one look at it and handed it back. "This stone is worth less than a setting," he said.

"That's impossible," I insisted. "Even in Brazil big gems aren't free. And she wouldn't have given me a fake!"

"It's not fake," Mr. Gare replied. "It's not glass or paste. It's real—corundum with traces of chromium. But it wasn't mined. It's synthetic."

He mounted my ruby on an optical instrument in the back of the store. I looked through the lens. I saw pure ruby red: what was I supposed to see? Defects, he replied, inclusions, dislocations. Every crystal pried from the earth betrays the pathologies of its history: mistakes in alignment, accidents in growth, deformations from pressures. Your crystal has none of these; it was grown in the utter peace of a laboratory. Only human intervention can create such perfection.

In a famous passage in *The Magic Mountain,* Thomas Mann's hero, observing snowflakes falling on his sleeve, shudders at their intricate, jewel-like precision:

Each in itself—this was the uncanny, the anti-organic, the life-denying character of them all—each of them was absolutely symmetrical; icily regular in form. They were too regular, as substance adapted to life never was to this degree—the living principle shuddered at this perfect precision, found it deathly, the very marrow of death.

I'd loved gem and mineral shows since childhood, but my interest was passive. The Gare-Mann debate lit a spark. Smith's science library was first rate; I could follow this up. I combed the shelves for books and articles on crystals. One footnote led to another and eventually across the campus to the art library circulation desk.

The Grammar of Ornament doesn't circulate, the librarian told me. I had to read it right there. That was just as well. Published in London in 1856, *The Grammar of Ornament* was the first book to be printed by chromolithography. "The drawing upon stone of the whole collection was entrusted to the care of Mr. Francis Bedford, who, with his able assistants...have executed the One Hundred Plates in less than one year," the author, Owen Jones, wrote in the preface. Lifting that gilt-edged, leather-bound, oversized and overweight tome, I suspected they'd left the stones in.

Jones and his assistants had carefully copied the ornamental patterns at their sources—a carved relief on the wall of an Indian temple, a tiled floor in a Byzantine palace, birds painted 5,000 years ago on the ceiling of an Egyptian tomb. Each of the hundred plates displays many different patterns, each is a collage of hexagons, triangles, squares, frets, lattices, strips and stripes, vines twisting right and left and right again. The region or the era in which they were found is written at the top of the plate: Egypt, Assyria and Persia, Greece, Pompeii, Rome, Italy, Byzantium, Arabia, Turkey, the Alhambra, India, Hindu, China. Celtic ornament, Medieval ornament, Renaissance, Elizabethan, the ornaments of savage tribes.

The thread I followed across the campus is this. In the science library, I had learned that the atomic pattern of any crystal is—or would be, if grown in the peace of a laboratory—as orderly as the field of stars on the American flag, as intricate as a Moghul rug, as repetitive as Victorian wallpaper. Crystal patterns are nature's designs. They follow the dictates of chemical bonds and thermodynamics. Yet, if left to grow undisturbed, these atomic motifs obey exactly, precisely, the same symmetries, the same rules of repetition, discovered and rediscovered by decorative artists in all times and places, from ancient Egypt to the Sandwich Islands. That means, the textbooks said, that the atoms, or molecules, in a crystal arrange themselves neatly in lines, the lines in rows, and the rows in planes, like a three-dimensional lattice.

Plate I, "Savage Tribes No. 1," Owen Jones, *The Grammar of Ornament* (London: Bernard and Quartich, 1868).

Jones compiled *The Grammar of Ornament* in the golden age of English encylopedias, but he did not call it an encyclopedia, nor did he mean it to be one. In those 100 plates, written in the visual languages of vastly different cultures, Jones sought, and found, a common grammar. "If the student will but endeavor to search out the thoughts which have been expressed in so many different languages," he wrote, "he may assuredly hope to find an ever-gushing fountain in place of a half-filled stagnant reservoir."

Owen Jones, architect, decorative artist, educator, and author, was born in London in 1809. Before writing his masterpiece, he traveled widely and published a major work on the fabulous Alhambra, the vast palace in Grenada, Spain, with no square inch undecorated. But it was an event at home that inspired the *Grammar*. As superintendent of works for the Great Exhibition of 1851, Jones's responsibilities included decorating the Crystal Palace and arranging the displays. Not only was he exposed to all the wealth of invention and design from all over the world, he designed the 14,000 showcases.

The Great Exhibition opened the eyes of its six million visitors, but Jones closed his in shame. He was embarrassed by the *inappropriateness* of European decorative art. There was no way around it: from deep antiquity to the present, the ornament of many non-European, even primitive, societies was far superior to that of civilization's supposed center. "Whilst in the works contributed by the various nations of Europe…there could be found but a fruitless struggle after novelty, irrespective of fitness," he wrote, "there were to be found in isolated collections…all the unity, all the truth, for which we had looked elsewhere in vain."

In all that geographical and temporal diversity, in that swirl of motifs and colors, Jones discerned four main principles. The best ornamental styles accord with form in nature. However varied ornament may seem, it follows but a very few simple laws. Change comes suddenly, by throwing off some fixed trammel. Future progress will blend tradition with a fresh look at nature. From these principles, Jones drew 35 precepts concerning the subservience of decorative arts to architecture, the preference of geometrical construction over lifelike rendering, and the uses of color. "See how various the forms, and how unvarying the principles," he wrote. "The works of the Creator are there to awaken…a desire to emulate in the works of our hands, the order, the symmetry, the grace, the fitness, which the Creator has sown broadcast over the earth."

Order and symmetry: just so. To mathematical eyes, the grammar of ornament is not *what* repeats, but *how*. As Jones pointed out—though he meant something else by it—the possibilities are few. By the end of the nineteenth century, mathematicians and crystallographers, together, had worked out the geometry and found all 230 symmetry types for patterns that repeat in three-dimensional space. And, I learned, *The Grammar of Ornament* was a grammar of symmetry, the richest compendium of two-dimensional repeating patterns in the world.

Seeing me pore over *The Grammar of Ornament* day after day, George Cohen, a soon-to-retire professor of art, wondered why. I told him.

"Do you know Dorothy Wrinch?" he asked.

"No," I said. "Who's she?"

"You should meet her," he said. "She owns a copy."

Dorothy had been at Smith even longer than he had. In 1939, when Hitler's armies crossed into Poland, she brought her young daughter from England to safety. After two years at Johns Hopkins in Baltimore, Dorothy married Otto Glaser, a biologist at Amherst College, and settled down in this valley of garden campuses, faint echoes of Oxford and Cambridge.

I climbed the stairs to her small office on the science center's fourth and top floor, across the hall from a suite of multipurpose labs, near the long narrow room that would house Smith's first computer.

Dorothy invited me in. She wore a flowered dress and cardigan sweater. She was slender, her hair was white, her eyes were blue. She was friendly, but all business. She didn't waste time on chitchat. No wonder I'd never seen her in the lounge.

The layout of her office was the same as mine two floors below. The floor-to-ceiling pane of dark gray glass dimmed the brilliant autumn sunlight, dulled the breathtaking leaves. Not a whiff of the seasons or even fresh air: the window was permanently sealed. Better this, sighed my department chairman when I complained, better this than the architects' first plan, which had no windows at all.[1]

My office was bare—just my desk, chair, and textbooks for my classes. Dorothy's was stuffed with a lifetime of scholarship. Journals, offprints, preprints. Books like *Dana's System of Mineralogy*, in two volumes; *Aspects of Form*; *Insect Architecture*; *Order and Life*; *Textiles Under the X-rays*; *Growth and Form*; *The Grammar of Ornament*, and three slim volumes she'd written herself. I remember their names because those books are mine now.

Books and models: I have her models too. Small, jewel-like, exquisitely beautiful skeletal models, made of pea-sized colored plastic balls joined by slender steel rods. Models of molecules, models of crystal structures, and the five regular polyhedra, the pinnacle of Euclid's thirteen books. These are the shapes Plato gave to the elements, earth, air, fire, and water (the fifth was the shape of the cosmos, he said). The shapes, Kepler discovered, that explain why God made only six planets. Elegant, symmetrical shapes that, like the Sirens of mythology, still bewitch us.

A box on a table held even smaller balls, stacked like cannonballs or oranges: these were nuggets of pearl tapioca. Dorothy had dyed them with food coloring and stuck them together with Elmer's glue to make a remarkable number of geometric shapes. You can buy the tapioca at State Street Fruit Store, she said. And so I did.

Dorothy reached into a glass case and handed me a curious metal cage, about four inches high, wide, and deep. It had been folded from a single metal sheet, a sheet perforated in a lacelike pattern of hexagons. "Niels Bohr made this for one me," she said. She'd made the other models herself.

Her desk was piled with papers. She was writing a book on crystal geometry.

Dorothy invited me to come again. I returned week after week. Once, as I was leaving, she asked me to make her a hexahedron.

A *what*? I asked, but then I understood. Four of the five regular poly-hedra, the tetrahedron, octahedron, dodecahedron, and icosahedron, are called by the Greek names for their number of facets: four, eight, twelve, and twenty. So why is the fifth called a *kybos*, or cube? Some scholars believe this linguistic anomaly shows the cube was discovered first, and the other four later.[2] But history need not be destiny. Nomenclature must be precise, said Dorothy. It should tell you what you need to know. "Cube" tells us nothing. Hexahedron henceforth.

Her request that I make a hexahedron was a test, I see that now. Not a test of whether I *could* build one, but whether I *would*. Had I dropped the "A"—had I asked *What? Make you a hexahedron?*—I would not be writing this book. But I knew even then that I'd found a problem that would engage my energies for the rest of my life: crystal geometry, pattern, growth, form. And I knew I'd found my teacher.

But I forgot about the hexahedron until the next time I headed up the stairs. I dashed back down to my office, cut out six squares of paper, and hastily taped them together. Plato would have frowned on the gaps between the squares, the crumpled corners, but my model captured the essence.

Dorothy took Plato's side. She'd asked for a hexahedron, not a mess. And so my education began. Of the many lessons I would learn in the years I assisted her, the most lasting is this: to understand any object, material or Platonic—to really understand it, to internalize it, to be able to use it—you must *build it yourself*. Draw projections, sections, perspectives. Make a model, and make it accurately, to scale. Take it apart, stir the pieces around, put them back together. Hold it in your hands, view it this way and that.

Back then, circa 1970, that lesson was mathematical heresy. From the end of World War II to the computer revolution, the dogma of abstraction ruled the known mathematical world. Geometers vied for status by claim-ing to never glance at a drawing. Dorothy could abstract with the best of them; later, I would learn that she had. But she also understood the Chinese proverb:

I hear and I forget,
I see and I remember,
I do and I understand.

A toolbox on her desk held a miniature hammer, a tiny pair of pliers, a small wrench, a small bottle of glue, and a vise. Alas, the manufacturer in Scotland refused to send her more precision-drilled colored plastic balls. No more steel rods either. She'd returned them one time too many. But she'd had to, hadn't she? she asked me. The balls must be drilled exactly to order, mustn't they? And the rods must be cut to just the right length.

We enjoyed each other's company. Both of us found Smith a lonely place. Dorothy seemed to know everyone, young and old, artists and musicians and writers. But her closest friends among the scientists had, all but Professor Soffer, in the course of time, died. And perhaps she saw her younger self in me and my tightrope act of family and career. But we never talked about that or the similarities between her first husband and the one who was mine at that time.

For me, Dorothy was an antidote to Smith's old-money diffidence. In that white-glove era, teachers and students alike took pride in their lack of ambition. In an admissions brochure that fanned my irritation to outrage, an eminent professor of English praised "gracious and thoughtful behavior" as the goal of a Smith education and noted, with evident satisfaction, "the relative absence among us of any misconceptions concerning our incomparable and far-reaching service to the world." I'd fled Kentucky at 17 to escape the gracious life. What was I doing at Smith?

The smugness would soon be gone with the wind (in the words of a Smith alumna). White gloves off, blue jeans on. Meanwhile, I treasured my hours with Dorothy.

Like every convert, I now found symmetry everywhere I looked, in everything I read. "God, thou great symmetry," wrote the poet Anna Wickham. Dorothy, I gathered, saw each in the other. Symmetry was a priori, the Beginning before it began. "Seek and ye shall find," Bach titled a canon in *A Musical Offering*. Seek: and you will find symmetry in the fundamental laws of physics, in a rainbow, in a snowflake, in a drop of water, in the glint of a crystal. In a clover, a pinecone, a blade of grass. In the weave of a cloth, the lace of a doily, in the stained-glass rosettes of Gothic cathedrals. In square dances and circle dances, in the *pas de quatre* of Swan Lake.

In what became a lifelong pattern, I taught myself by teaching. Ergo, my course on symmetry. More cautious colleagues doubted I knew enough about

science, music, art, dance, and literature to lead students through these labyrinths. I knew I didn't, but so what? The course would be an adventure: my
students and I would explore the labyrinths together. We couldn't get lost—I
held Ariadne's thread, mathematics, the passkey to everything. What do a
molecule, a circle dance, an apple blossom, a kaleidoscope have in common?
Symmetries: mirror reflections, circular motions. Add, to these two, marching in line, twisting in a helix, and reversing antipodes, and you know all
you need to know. The five motions of symmetry are the fundamentals of all
repeating patterns, the laws behind the laws of ornaments and crystals alike,
simplicity beneath profusion.

You call this a math course? one colleague muttered. But the art will draw
students, another replied. They sent it on to the college's committee on new
courses, chaired by the dean. The dean, Phyllis Williams Lehmann, the eminent
archaeologist and art historian. The dean whose imposing presence, brilliant
mind, and instantaneous response time had so intimidated me at faculty meetings, even unto the very back row. When my proposal reached her desk, Dean
Lehmann scanned it with her eagle eyes and reached for the phone. "Surely you
do not *presume* to teach *art*!" she said. Said, not asked; it wasn't a question.

I'd put *The Grammar of Ornament* on the reading list.

Scientists seized on *The Grammar of Ornament* the moment it appeared.
The most famous physicist of the time, James Clerk Maxwell, consulted it in
designing a tiled floor for his family home in Scotland. Maxwell was a student of color, as well as of electricity and magnetism; he won the Rumford
Medal for his work on color perception and color blindness, and was an early
pioneer of color photography. In designing the floor, Maxwell studied the
colors in *The Grammar of Ornament*, and Jones's precepts for using them,
with great care. Maxwell had a kaleidoscope too.

Owen Jones, James Maxwell, and the kaleidoscope's inventor, David
Brewster, knew of one another. Did they ever meet? I hope they did; they had
a lot to talk about. All three cared deeply about ornament and its grammar.
Brewster may seem the odd man out to us, but *he* would not have thought
so, nor would they. Brewster never meant his "instrument," as he called it, to
be a toy, though he allowed as it might moonlight in that role. He invented
the kaleidoscope in 1815, he said, as a precision tool for artists to "employ in
the numerous branches of the useful and ornamental arts," like stained-glass
windows, architectural ornaments, the decoration of public halls and galler-
ies, carpets, and jewelry. He said precision and he meant it; the vagueness of
"kaleidoscopic" would have made him sick.

Brewster invented all the variations you find in craft shops today, and
many more: two-mirror and three-mirror kaleidoscopes, of course, and also

3-D kaleidoscopes, stereoscopic kaleidoscopes, microscopic kaleidoscopes, kaleidoscope cameras, and "kaleidoscopes which combine the colours and forms produced by polarized light." He urged artisans to use these marvelous instruments for the purpose for which he'd designed them. But it seems they paid him no heed.

Why not? The art historian E. H. Gombrich replied before I asked. "The idea of a pure form is always an artificial abstraction. The rose-window in a Gothic cathedral is not the result of chance produced in a pretty toy. We cannot divorce the impression it makes from our knowledge of the craftsman's skill and from awareness of meaning."[3]

But I did not presume to teach art. I knew—or so I thought then—all I needed to know. James Clerk Maxwell and Dorothy Wrinch and I and my students could, and happily did, divorce pattern from meaning. We read the visual languages our way.

The kaleidoscope works its magic through mirrors. The simplest kind uses just two. Two mirrors, and tiny bits of something, perhaps colored glass: that's all there is, but that's not all we see. The beauty lies, not in the tiny bits, but in their repeated, multiple, reflections. You don't need a kaleidoscope tube to see this effect; just tape two pocket mirrors together along an edge and put those tiny bits in the wedge between them. The reflected images will form a perfectly symmetrical rosette, Brewster taught, if you choose the angle between the mirrors so that the number of mirror-images pairs divides the circle in equal parts. Since one mirror-image pair spans twice the angle between the mirrors, the angle itself must divide 180 degrees. For example, if the angle is 60 degrees (as in figure 1) you will get a rosette with three mirror-image pairs, since 180/60 = 3.

A two-mirror kaleidoscope makes a circular pattern. If you close the wedge formed by the mirrors with a third mirror, making a triangular prism, the third mirror reflects the circular pattern as a whole, and that reflection is reflected in the other mirrors, ad infinitum; the pattern repeats ornamentally forever. Maxwell needed three mirrors to design his floor.

Figure 1 With one mirror, we see an object and its mirror image. With two mirrors, the object reflects back and forth to give the illusion of a rosette.

Figure 2 In principle, a pattern generated by reflections in three mirrors continues forever in all directions. Here the mirrors form an equilateral triangle.

In a three-mirror kaleidoscope, the condition that the angle between the mirrors must divide 180 has to hold at each of the three angles of the triangular prism. And these angles together are subject to a constraint: since the three mirrors form a triangle, the three angles must add up to 180 degrees. These joint conditions—that each angle divides 180 and the three angles add up to 180—are very restrictive. There are only three solutions: the prism can be a 30–60–90 triangle, an isosceles right triangle, or an equilateral triangle. Only these three. (See figure 2.)

And that is why, in the delightful swirl of motifs and colors that a three-mirror kaleidoscope presents to our eyes, the underlying pattern can only be hexagons or squares or triangles. No pentagons or any other -gons. "See how various the forms, and how unvarying the principles!" All three kaleidoscope types can be found in almost every plate of *The Grammar of Ornament*.

Dean Lehmann changed "ornamental art" in my proposal to "ornamental patterns" and said okay.

Wall posters in the science center's narrow corridors announced the visit of the famous chemist and double Nobel laureate Linus Pauling. Innocently, I asked Dorothy if she planned to go to his lecture and the dinner for faculty afterward. "No," she said. End of conversation.

I went. Pauling talked about crystal patterns. Think of the atoms as spheres, he said, and the patterns as spheres packed and stacked. Some of these arrangements are lattice-like, some are not. I nodded happily. From my study of *The Grammar of Ornament*, I could spot the five planar lattices a mile away. Every repeating pattern in the plane, kaleidoscopic patterns included, is built on one of those five. Though three-dimensional lattices are more complicated, I could visualize them, more or less. But I could not imagine the others, the nonlattice packings. What did they look like? After dinner, I marshaled my courage and asked him. The great man stared at me for a second or two, then turned to his wife. "Where's my coat?" he asked her, and they walked away. Years later, at the end of his life, he would grapple with my question.

In the coffee lounge, not long after Pauling's visit, Professor Soffer looked up from his newspaper again to ask, "You spend a lot of time with Dorothy Wrinch. How much do you know about her?"

"Not much," I said. "She never talks about herself."

"You should know," he said, and sketched her story. Dorothy, a mathematician, had traipsed into protein chemistry in the 1930s. The model she came up with for protein molecules set off a buried land mine. Linus Pauling claimed it couldn't exist and hounded her out of the field. She found safe harbor at Smith and fought back from here, but no one listened. In 1952, a key element of her structure was found after all, but not in proteins. Until he, Professor Soffer, came across the paper in a journal two years later, no one had told her.

Dorothy told me later, "First they said my structure could not exist in nature. Then when it was found in nature, they said it couldn't be synthesized. Then when it was synthesized, they said it wasn't important anyway."

"Among the artists, as among the scientists, there are some who see beauty and significance in symmetry," Dorothy wrote in a polemical press release I found among her papers. "The Greeks belong in this category. So do Einstein, Irving Langmuir, and Niels Bohr." To which she added bitterly, "There are others to whom such inexorability and order mean nothing. Many scientists of yesterday and some of today belong here."

In 1973, I organized a Symmetry Festival with colleagues in chemistry and geology. The three-day celebration included change-ringing on handbells, a concert of Renaissance dance followed by lessons for the audience, several exhibits, lectures on typographic ornament, color and symmetry, and ambiguities in symmetry-seeking with emphasis on the stories of Borges. The Smith music department performed Bach's "Musical Offering." My daughter Diana, in grade school by then, made symmetrically shaped cookies for a coffee hour.[4] In an essay for the Festival guidebook, I wrote:

> *The theory of symmetry is a triumph of the human intellect. It is the perception of order in a chaotic universe, the study of the forms that order can take, and the use of that study to give significance to the things we see. In science as well as in the arts, symmetry is the geometric plan on which the variations of nature and of life are drawn.*

I believed that, then.

4 DOT

The Wrinch papers, 30 burgeoning boxes, reveal little of Dorothy's first 35 years. But the Internet is a real net, culling long-lost details from the depths: nineteenth-century census records, marriage and baptism records in an Argentine Anglican church. I've found her school's Jubilee book and letters in far-flung archives, and family members alive and well and on Facebook.

"Grandma was a tough cookie," says Talitha Williams, Dorothy's niece.

Grandma—Dorothy's mother, Ada Minnie Souter—was born in 1867, near Bury St. Edmunds, the Suffolk market town where her parents grew up. She left home at 26, as her father was ending his long career as station master for the Great Eastern Railway. Robert Souter's last posting was the same as his first, the small town of Geldeston on the Waveney River. The pace was semiretired: six or seven trains rolled through daily each way. He kept the books, ran the station, refreshed train crews, assisted passengers, loaded grain, herded cattle into designated carriages, sent drunks home, and shooed dogs and children off the tracks.

It took a tough cookie to quit her job as school headmistress and set off for Argentina alone.[1] Her fiancé, Hugh Edward Hart Wrinch, had gone there three years before. A year older, Hugh was born and raised in not-far-away Woodbridge, on the River Deben. The town had once been a busy port, a center of shipbuilding, rope-making, and sail-making. Steamships and railroads were putting the ancient trades out of business, but ships still arrived in Woodbridge by the hundreds, bearing coal in trade for corn, flour, and malt from the rich agricultural hinterlands.[2]

Hugh Wrinch was the fourth son of a corn, coal, and malt merchant. The family had been well-to-do: the 1861 census lists a household of two parents, several children, a maiden aunt (the "fundholder"), and three live-in servants. Ten years later, when

Hugh was five, there were more children and fewer servants. A severe agricultural depression had devastated farmers, merchants, and shipping firms alike. Young men were leaving the countryside and the country in droves, in search of better lives and livelihoods.

Best to take skills with you when you go. Hugh left the Woodbridge Grammar School at 17 to apprentice with the Messrs. Whitmore & Binyon in nearby Wickham Market, a firm renowned for its steam engines, water wheels, windmills, corn and flour mills. He spent five years at the lathe, at the fitting bench, in the erecting shops, and in the drawing office, and then two years in the Great Eastern Railway's Locomotive Department at Stratford repairing, maintaining, troubleshooting, and testing.[3] Farther, faster, better: Great Eastern could scarcely keep pace with demand. Children chanted:

Faster than fairies, faster than witches,
Bridges and houses, hedges and ditches,
And charging along like troops in a battle
All through the meadows the horses and cattle:
All of the sights of the hill and the plain
Fly as thick as driving rain;
And ever again, in the wink of an eye,
Painted stations whistle by.[4]

By 1890, Hugh was ready to leave. "It gives me much pleasure to say that he is a very good and intelligent workman," wrote his railyard supervisor, "and during the time he has been in this service has conducted himself most satisfactorily." Whitmore concurred. The lad is "possessed of more than ordinary abilities, and has plenty of go in him. [We] believe him to be in every respect trustworthy and competent to take a leading position in an Engineering Firm abroad, as he is desirous to do."

In 1890, British investment in Argentina totaled 157 million pounds (over 16 billion in current pounds);[5] projects included railways with 5,800 miles of track and a major expansion of the waterworks. Immigration had tripled the Argentine population in just two decades. In that rapidly developing outpost of the British commercial empire, newcomers found work in an alphabet of skilled trades from accounting to upholstering.

Hugh found work with the Consolidated Water Works Company (Aguas Corrientes) in Rosario, a thriving city on the Paraná River 187 miles northwest of Buenos Aires. Soon he was promoted to pumping station manager.

Mail boats brought the *Woodbridge Reporter & Wickham Market Gazette* with latest news from home.[6] In September 1892—springtime in Rosario—Suffolk roadsides were "beginning to present a very pretty appearance with tinted leaves inter-mixed with blackberries which are very plentiful this year," and a donkey driver had "committed an offence against the bye-laws of the Felixstowe Local Board, by driving more than three asses at one time." The next September, the mail boat brought Ada Minnie Souter. They were married in St. Bartholemew's Anglican Church on November 11.[7] Their first child was born eleven months later, on September 13, 1894. They baptized her Dorothy Maud and called her Dot.

Few homes in Rosario (or in Buenos Aires) had running water when Hugh arrived in 1890; by 1897, Queen Victoria's Diamond Jubilee, two million gallons reached the town every day. The company rewarded him handsomely, but he resigned for unspecified health reasons that same year. The family returned to England, and Hugh was promptly hired by the Chelsea Waterworks Company, Ltd.

Chelsea, founded in 1722 by an act of Parliament, was one of the oldest private companies pumping water to London, and one of the most highly regarded. It had introduced slow sand filtration before the cholera epidemic of 1854, a public health measure that proved its worth when Chelsea customers—including the royal palaces—largely escaped the horror. When the epidemic subsided, the company relocated upstream at Seething Wells in Surbiton and built new reservoirs and filter beds. Hugh was put in charge of the Surbiton plant.

"As a riverside resort [Surbiton] has always been popular," says a 1905 guidebook to London. "There are promenades along the river bank, well laid out, lined with fine old elm trees, and provided with seats; a band plays twice a week during the summer."[8] Surbiton, sniffed Virginia Woolf in the pages of *Good Housekeeping*,[9] is where London's shop assistants live, with their "gramophones and wireless, and money to spend at the movies." The Wrinch family home was two-story red brick with a garden, on Portsmouth Road along the Thames. Their second child, Muriel Louise, was born in 1899.

Today, the reservoirs and filter beds are home to wild waterfowl. The old, brown brick buildings, woven incongruously into the new campus of Kingston College, were once the waterworks plant: the gate house, with Hugh Wrinch's office, is now the university porters' lodge; the two-story social club for engine drivers, cleaners, stokers, and yardmen is a residence hall; and the maintenance store, coal store, and muniments (deeds and documents) buildings stand empty. The old pumping station is a fitness center; of the magnificent steam engines with enormous wheels that dwarfed and deafened the workers there is no trace. They were works of art, with brightly painted metal, gleaming brass fittings, and polished casings of burnished wood. Cleaners and polishers worked night and day to keep them running at maximum efficiency.[10]

Dot entered Surbiton High School in January 1899, at the age of four and a third. The school, for girls from tykes to teens, had struggled since the Church Schools Company founded it 15 years before, and Alice Maud Procter, the visionary new headmistress, was trawling for students. Archdeacon Burney of St. Mark's up the hill gave Miss Procter leave to make changes, but she had to raise the money. Eliza Burney, his daughter, would be her right hand.

The previous headmistress had left to do missionary work in Africa; Miss Procter found her mission under her nose. The school was small, just two shabby buildings, and the walls were bare. Children shivered in wooden chairs attached to desks, facing front; the fires in the back warmed only the chimneys. A tiny garden doubled as a playground. There were no art or science teachers. In assembly each day, the students sang—*sang!*—the school motto: "Be good, sweet maid, and let who will be clever."[11]

In her first year, Miss Procter drew up plans for a new school wing; introduced the School Litany and Second Prayers; started a choir and an Old Girls' network; and cajoled staff, parents, and friends to donate pictures to brighten things up. She hired a science teacher, a modern language teacher, a geography teacher, and a music teacher and dispatched the math teacher to lectures on new teaching methods. To cover these costs, she recruited new students; tuition ran four to eighteen guineas.

Shabby housing, low standards, a weak curriculum, and stultifying motto notwithstanding, Surbiton High School had a "thriving culture of competition and rewards." This, said Miss Procter, had to go, *right away*: it devastated the children and poisoned the atmosphere. She replaced the school motto with "What touches one, touches all" and abolished marks and prizes. Competition only on the playing field! Except there was no field. Three-legged, egg-and-spoon, sack, threading the needle, high jump, long jump, and slow bicycle races were held in the garden until Dot's father, Hugh Wrinch, Honorary Secretary for the Games Field, leased space from the Water Works. "'Service' and 'fellowship' sum up the ideals I set before myself," Miss Procter said on her retirement 34 years later. "Each member of the school is as important as every other member, from the youngest to the oldest, forming one whole, linked together by love and a common purpose."[12]

The old buildings still stand today; the school has grown up around them. Dot stares out from an undated photograph in a historical display in the spacious, glass-walled lobby. An unnamed teacher—not Miss Procter—sits surrounded by seven small unnamed girls with straw bonnets. Dot is unmistakably the second

Students and teacher, Surbiton High School, circa 1902. Dot Wrinch is second from the left. Historical display, Surbiton High School, 2009.

from the left, the one with flowers pinned to her white dress. She looks about eight; that would be 1902, the year the town built a clock tower near the school to honor the coronation of King Edward VII.

From childhood, Dot and Muriel were guided—no, ruled—by not one but two tough cookies. Miss Procter was the other. She had fought her way loose from a cosseted life in India: homeschooled, waited on hand and foot. When she was nine, her parents brought their four daughters to England on home leave and she met her Procter kin for the first time.[13] Poets ran in the family: Aunt Adelaide was Queen Victoria's favorite. Adelaide's most-quoted line, often misattributed to George Eliot, was, "We may always be what we might have been."[14]

Adelaide died at 38, too young to meet her nieces, but her friends remained in the Procter circle. They were passionate supporters of a bold new idea: higher education for women. They quoted their friend Lord Tennyson's "The Princess":

O I wish

> That I were some great princess, I would build
> Far off from men a college like a man's,
> And I would teach them all that men are taught...[15]

One of Adelaide's friends had done just that.[16] She set her college in Hitchen, 21 long miles from Cambridge. That proved to be too far off; within a few years it was moved to Girton.

When their parents returned to India, Alice Procter and her younger sister Zoë stayed behind in England with a woman who took in "Indian children" for a fee. Mrs. Wilson was an aunt of Robert Louis Stevenson; in her autobiography, Zoë recalled sitting on his knee. Mrs. Wilson sent the girls to a school in London housed in a mansion that had belonged to Captain Cook. The great explorer had died a century before, but, the girls whispered, he stalked the top floor every night, his telescope under his arm.

The school was run by the Girls' Public Day School Company (GPDSC), a new experiment in girls' education.[17] Its motto, "No more is knowledge a sealed fountain," was also a line from "The Princess." GPDSC girls didn't sing or embroider their motto—they studied for Oxford and Cambridge.

Personal mottos were popular then, like e-mail signatures today; Zoë Procter chose, "What man has done, woman can do." A classmate, Philippa Fawcett, already brilliant at math, would later prove her right, but Philippa's mother Millicent inspired more awe at the time. Mrs. Fawcett was a leader of the campaign to give women the vote and had helped found Newnham, Cambridge's second college for women. Zoë proudly presented Mrs. Fawcett with a bouquet at a prize ceremony.

Alice Procter dreamed of becoming a teacher. The path was neither easy nor straight, but Aunt Adelaide's friends urged her to become what she might have been. At last, at the age of 24, she went up to Girton.

Alice, with her patchy education, was not Girton's leading scholar, but her third-class honors in history got her a teaching post at a Girls' Public Day School in Bristol. Then, after seven years, "impelled by some driving force," she applied for the vacancy at Surbiton.

Miss Procter's role model was Thomas Arnold, the legendary headmaster of Rugby School. On taking up his duties there in 1828, the Great Educator had vowed to "change the face of education all through the public schools of England." In the 13 years of life left to him, he did it. God moved to center stage: "First religious and moral principles, second gentlemanly conduct, third academic ability." To co-opt bullies, Arnold transformed the Sixth Form into "symbols of moral and academic integrity" responsible for school discipline and responsible directly to him. In fact, the bullying never stopped, and younger boys still "fagged" for the older ones, blacking their boots, brushing their clothes, running errands, even warming their toilet seats. But

Dr. Arnold may not have noticed. Buffered by teachers, Sixth Formers and lofty principles, he ruled from an airy office at the top of a high spiral staircase that boys climbed when, and only when, summoned. For Arnold, the much-admired headmaster of Tom Brown's school days, was also the terrifying headmaster of *Eminent Victorians*. "Among the lower forms of the school his appearances were rare and transitory, and upon these young children 'the chief impression,' we are told, 'was of extreme fear.'"[18]

Thomas Arnold stalked the halls of Surbiton High School as resolutely as Captain Cook stalked his mansion, but with a Bible instead of a telescope. "One memory alone" remained "sharply clear from my first day until the morning I waited to sing 'Lord dismiss us with Thy blessing' for the last time," one of Dot Wrinch's classmates recalled years later, "and that is the memory of wooden heels on wooden floors, heard far off and then growing nearer—the click-click that told us Miss Procter was coming."

Miss Procter saw hell's flames in every recess of character. "For the sake of your souls, it would be a pity to use your influence for evil when you might use it for good," she admonished devilish Sixth Formers.[19] "Very clearly were our responsibilities put before us by the Head Mistress and our failings laid bare," another Old Girl wrote. "I remember one occasion when the Sixth was sent for to the Office in a body; our joint delinquency has faded from my mind, but the horror of that moment I shall never, *never* forget." But, she added in her own defense, "It was not easy to play the heavy senior to the then irrepressible Dorothy Wrinch."

But Miss Procter could be kind, beyond the call of Higher Duty. One darkening afternoon, when a girl's front tooth was knocked out in a game, the headmistress lit matches and led the staff and Thomas the cat around the yard in a fruitless search for it.

Sixty years later, in a time of depression and despair, Dorothy had forgotten Thomas and the tooth, but not a dreaded session in the Office. *Well, all I have to say, my dear little Dorothy, is that you will come to a bad end.* Dorothy mulled over those words as from an oracle. "Proc's venom and hostility was crucial for me, in that she could withdraw her support, through which alone I could get to Cambridge…and I wanted to get to Cambridge."[20]

The self-described "wee little thing" complained to her parents, but they could do nothing to help her. Or they chose not to. Ada Minnie expected her daughters to cope, to take responsibility for their actions. "My mother had brown hair and wanted to be a blond," says Dot's niece Talitha. "She tried to dye it with peroxide but it came out streaked. Grandma made her wear it that way until it grew out."

Zoë Procter was an admired and influential "presence" in her sister's school. By 1912, the year Dot graduated, Zoë and Alice had parted ways with Mrs. Fawcett and her nonviolent tactics in the suffrage struggle. The peaceful century-long campaign to give women the vote was going nowhere; Parliament still ignored their petitions. *Deeds, not Words*! cried Christabel Pankhurst, a leader of the movement's most radical wing, calling for an evening of "symbolic" acts of violence in London. Zoë signed up. On March 12, she hid a hammer in her muff and stood on alert in St. Martins-in-the Fields. "I heard the clock strike six…immediately afterwards there sounded the crash of shattered glass from the direction of the Cunard Company's office, and I hastily turned the corner and swung my hammer against several of the small panes of an old-fashioned silversmith's shop." An uncle hired a lawyer to defend her, but Zoë refused to cooperate and proudly went to jail. Alice Procter brought her sister baskets of fruit. Dot would rebel against the church, and the team spirit failed to take root. But the Procter sisters imbued her with their passion for women's rights.

That same spring, Dot applied to Girton, presenting certificates in History, English, Mechanics, French, a paper on Latin, and Mathematics (elementary), Mathematics (advanced). She had also studied Mathematics at University College, London. "Proc" came through for her after all, with a glowing recommendation, and the Canon of St. Marks vouched for the Wrinch family's respectability. Eliza Burney, the Archdeacon's daughter, still Miss Procter's right hand, was more cautious: "I think her a little too much inclined to sentimentalism, and naturally sensitive…but she's a dear girl and thoroughly conscientious—I feel nearly sure that her influence with her companions was to the good."[21] Dot deferred college for a year, probably to prepare for the rigors ahead, and went up to Girton in the fall of 1913.

II LOGICS

Or if this all which round about we see,
As idle Morpheus some sick brains have taught,
Of undivided motes compacted be,
How was this goodly architecture wrought?
Or by what means were they together brought?
They err that say they did concur by chance;
Love made them meet in a well-ordered dance.
"Orchestra, or a poem on dancing," Stanza 20

—SIR JOHN DAVIES, 1596

5 THE WRANGLER

Girton College was tugged into the world by many capable mid-wives, Bertrand Russell's grandmother among them. But Girton's mother, if that word can be applied to the most unmotherly of women, was Miss Emily Davies, an abrasive, indefatigable, indomitable fighter for women's higher education. Brooking no concession to their supposedly different interests or limited career prospects, Miss Davies demanded, and got, a curriculum as rigorous as that of ultra-rigorous Trinity College in Cambridge. She also demanded, but did not get, university recognition for the new college. After all, membership in the university (which conferred degrees) carried with it the right to vote on university matters. Since women did not have the vote, granting full degrees would be oxymoronic. Worse, members of the university could borrow books from its library! Cambridge would not fully recognize its women's colleges until 1948.

But Cambridge did admit women students to university exams. Miss Davies chaperoned hers to Cambridge for the purpose and sat with them the while, knitting by the fire.[1] Throughout her reign and long afterward, Girton students were chaperoned everywhere, even to afternoon teas.

"Contrary to the current opinion we Girton girls did not wear severe shirt blouses with formal ties, nor did we drag back our hair," said Dot's friend Dora Black, but that's how they dressed for the entering class photographs. Dot went up, a year after Dora, with a three-year College Scholarship for mathematics of £50. She was one of two scholarship students that year; the others paid their own way.

Identical curricula notwithstanding, Girton College *looked* nothing like Trinity. Trinity is a graceful poem in stone, stained

Girton College, entering class, 1913. Dorothy Wrinch is second from left in the first full row. By permission of The Mistress and Fellows, Girton College, Cambridge.

glass, and ivy. Girton, a sprawling red brick edifice, resembled (said C. S. Lewis) the Castle of Otranto, the eponymous setting for Horace Walpole's gothic novel.[2] Girton could have been its film set: an Old Girl, Hertha Ayrton, patented the first steady electric arc lamp in 1913; at last, cinematographers could film on location. But Lenin would electrify Russia before Girton replaced gas lighting.

Girton's dim interior was suitably medieval. Portraits of the formidable founders stared stonily onto cold stone floors. One dormitory wing had no baths. Some daughters of wealthy families had never lit a fire or boiled an egg; one had never washed her own hair. And the food! In Virginia Woolf's disputed description in *A Room of One's Own*, even High Table at Girton was revolting.

A horrified mother begged her daughter to come home. The daughter refused: at Girton each student had not one but *two* rooms of her own, one for study, one for sleeping, and a fireplace with a copper kettle. In rooms of her own, a girl could do whatever she liked except entertain men. No two rooms were exactly alike, not in size, doors, or windows, and "those who remember the little corners, the Gothic arches, the turret staircase, the queer little eccentricities of some especial corner will feel a special devotion that resists all change."[3] The rooms flanked corridors, not staircase landings as in the men's colleges, and chatter bounced up and down the halls. Servants (called gyps) brought jugs of fresh milk to the girls' rooms each evening, the better for study. But even the driven, workaholic Dot joined in late-night cocoa parties and singing.

Despite the chaperones (who sometimes winked), Girton meant freedom: freedom from parents, Miss Procters, and prayers; freedom from customs, conventions, expectations. As the years passed, memories would soften the contours of the castle, brighten the lights, ease the chill. Dot left a thousand dollars to Girton in her will, but the college never received it.

At his death in 1970, at the age of 97, Bertrand Russell was fading from public memory. Forty years later the graphic biography, *Logicomix*, brought him back. Here he is again, handsome and gaunt, in living color, Dot's college hero, the great logician. And my college hero too, the later Russell, steadfastly campaigning against the bomb and working tirelessly for world peace, Russell the Nobel laureate, Russell Nehru's friend.[4]

"Bertrand Russell is a hard man to do justice to," the *Logicomix* authors wrote, thanking the Bertrand Russell Society for its book prize. "He is too large to be contained in three hundred and fifty pages of words and pictures, too complex to be narrowed down to formulas of explanation, too interesting to satisfy the instincts of the storyteller for clear lines. But he was certainly worth the time and the trouble, Bertie was."

Dot first heard him lecture at Cambridge University on January 31, 1914. The hall was packed, standing room only. "To the students, Mr. Russell was an almost superhuman person," one of them wrote later. "I cannot adequately describe the respect, adoration, and even awe which he inspired."[5] Charles Kay Ogden, the young editor and publisher of *Cambridge Magazine*, had announced the lecture in its pages and reserved a double room. Dot sat somewhere in the back, anonymous like the other female students in her voluminous neck-to-toe, wide-sleeved black academic gown, the college-issue burqa. She must have come by horse-drawn cab: January days are short and cold, too dark and too cold to walk or bicycle the two miles from Girton to Cambridge.

Russell's title, "Our Knowledge of the External World," was timely. The external world was rapidly shrinking. That very month a single-passenger air-boat cobbled from spruce, cloth, and wire had crossed Tampa Bay in 23 minutes, the world's first scheduled flight. The first steamboat passed through the Panama Canal. Model-T Fords rolled off Detroit assembly lines in just an hour and a half, from start to finish.

But Russell hadn't come to talk about travel. He'd planned his lecture for serious philosophers, the few who really cared: cared how, from the colors, sounds, tastes, and smells accessible through our senses, we infer the existence of a world outside ourselves.

But a crowd had come. What should he do?

Make them laugh: "Philosophy, from the earliest times, has made greater claims, and achieved fewer results, than any other branch of learning. Ever since Thales said that all is water, philosophers have been ready with glib assertions about the sum-total of things…I believe that the time has now arrived when this unsatisfactory state of things can be brought to an end." The audience roared, and Russell relaxed.

He needn't have worried; it didn't matter what he said. The students, fresh from church-centric schooling, warnings of hellfire, brimstone, and utter darkness still ringing in their ears, had come to worship their hero. Russell was their Moses, parting the waters and leading them out of Egypt. "What a relief it was to hear [his] crack that the ten commandments were like the customary rubric for a ten-question examination paper: 'only six need be attempted.'"

If Russell was their Moses, Og—C. K. Ogden to everyone but Dot and Dora—was, in those years, their Aaron, their publicist and organizer. On Sunday nights, students from all 23 Cambridge colleges, Girton and Newnham included, squeezed into his cluttered rooms above a fish shop to debate philosophy, art, and literature. Og and a few friends had started "The Heretics" a few years before. The club was a response to a response to a provocation. The chain reaction began when the master of Emmanuel College wrote an essay with a shocking title, "Prove all things; hold fast that which is good" (Thessalonians 5:21). Proper Cambridge dons were outraged. "I think it a distinct abuse of your position," wrote one. "You have no moral right to upset a boy's religious beliefs unless you have a higher religion to offer him." But the boys were not upset at all. Pledging to reject all appeal to Authority, they formed the club to debate all things. Russell and other dons who cheered them on were made Honorary Heretics.

Dora had never met a nonbeliever before she came to Girton, but she was one herself before the year was out. Dot hesitated less. Dora and Dot were Heretics regulars. Girton's headmistress, Miss E. E. Constance Jones, was theoretically opposed but didn't ground them. Miss Jones ruled by example, not diktat. She opened each day with hymns, psalms, and prayers in the chapel before breakfast, but never took attendance. Nonconformist, atheist, and Jewish students had enrolled in Girton from the start. Hertha Ayrton, the electrical engineer (and George Eliot's model for Mira in *Daniel Deronda*), was its first Jewish student. Even before it opened in Hitchin, the college was known as "that infidel place."

"We didn't discuss politics," wrote Dora, "except votes for women." In her first year at Girton, Dot played on the tennis team and joined the Society for the Study of Little-Known Literature, the Musical Society, and the Mathematical Club, founded in Girton's earliest days.[6]

There was no Society for Scandalous Literature, but the girls kept up with the times. "The book for us girls was *Ann Veronica*." They knew the story was true, and they knew of whom it was true. Its married, middle-aged author, H. G. Wells, had seduced Amber Reeves, a brilliant Newnham student, in her college rooms! And then he wrote up their adventure, roman à clef style! Proper English were horrified, but Girtonians were impressed and inspired. Amber refused to give up her baby. Nor did she hide in shame and disgrace. The principals lived more or less happily ever after, on the page and in life. Dora longed to play Ann Veronica in real life and, after college, did it quite well. As Mrs. Bertrand Russell, she even had an affair with H. G.,[7] but it was brief, as was her husband's with Amber.[8]

Dot went up to Girton to read math; she read day and night. To become a mathematician (whatever she thought that meant—no profession is so invisible to its students) she had to do well—very, very well—in the marathon Mathematical Tripos.

The curious term "Tripos" harks back hundreds of years to the days when university examiners sat on tripod stools in the vast Senate House and read questions aloud. The students stood and wrangled with their interlocutors. "Tripos" eventually came to mean both a Cambridge course of study and the grueling examination at the end of it. The first part of the Mathematical Tripos exam was, and still is, taken by all mathematics students at the end of their first or second year. Students seeking honors took, and still take, Part II at the conclusion of their third. The top 30 scorers on Part II of the Mathematical Tripos are called Wranglers (for all other Triposes, the term "First Class Honours" is used). Those who didn't make (and some who did) changed fields.

Until 1910, the Wranglers were ranked by their Tripos scores: Senior Wrangler, Second Wrangler, Third Wrangler, and so on down to 30th. Just as your IQ once pegged you for life, the Mathematical Tripos in its heyday was thought to be the exact and immutable measure of each and every man who took it, from the moment he turned in his examination booklet to the end of his days on earth. Ambitious contenders trained for years and towns-people bet on the favorites. The results were announced, to deafening cheers, in the Cambridge Senate House on a May morning at the stroke of ten. The top three Wranglers were instant celebrities, feted like sports heroes, lionized by the press, photographed for postcards.

Since Girton and Newnham were unofficial colleges, the women's scores on the exams were unofficial too. But the ranks they *would* have held were announced verbally, along with the men's. Girton's Charlotte Angas Scott tied for eighth place in 1880, allaying the fears of doctors and parents that women, like quite a few of the men, would crack under the stress. Ten years later, Phillipa Fawcett of Newnham College scored *above* first place. Newnham's parades and bonfires lasted far into the night.

"The English have always had more faith in competitive examinations than any other people (except perhaps the Imperial Chinese)," said C. P. Snow.[9] The articles of faith in the Tripos were two: first, that mathematics was "the ideal foundation…to educate the future leaders of Britain and her Empire," and second, that this grueling test of mental agility measured the qualities needed to shoulder this burden. In the words of William Whewell, the influential master of Trinity for 25 years from 1841 until he fell from his horse:[10]

There is no study by which the Reason can be so exactly and rigorously exercised. In learning Geometry the student is rendered familiar with the most perfect examples of strict inference…he learns continuity of attention, coherency of thought, and confidence in the power of human Reason to arrive at the truth.

This faith was, of course, self-fulfilling. The long list of Senior and Second Wranglers includes four lord justices of appeal, four prominent actuaries, two bishops, an archdeacon, a president of the General Medical Council (who spoke 18 foreign languages fluently), a deputy speaker of the House of Commons, and many famous physicists.

Many famous physicists, but few world-class mathematicians. That's because, the Cambridge mathematician G. H. Hardy (Fourth Wrangler, 1898) insisted on every possible occasion, the Mathematical Tripos didn't measure aptitude for creative work in mathematics; no exam could. Students spent years learning useless clever tricks, never glimpsing the true frontiers of their subject. The Tripos had frozen the curriculum and retarded the development of British mathematics for 150 years. Abolish it! he thundered. Reformers won a partial victory: from 1910 and forever after, the Wranglers were listed alphabetically.

By 1914, when Dot took Part I, the public frenzy had abated a bit. But Wranglerhood was still essential for a mathematical career. And the winds of change had scarcely touched the Tripos's content. Nothing that Russell taught was on it.

After Newton and Leibniz's bitter squabble over priority for the calculus, English and continental mathematics went their separate ways. Gradually, like finches on the Galápagos, they acquired distinctive features and niches. Cambridge University, the uncontested mathematical center of all Britain, became a hub of mathematical physics. George Airy measured the mean density of the earth and established Greenwich as the prime meridian; James Clerk Maxwell unified electricity and magnetism; George Stokes made fundamental contributions to fluid dynamics; Lord Kelvin found the first and second laws of thermodynamics, defined "Absolute Zero," and laid the first transatlantic cable. The chancellor of Cambridge University in 1913, the year Dot came up, was Lord Rayleigh, author of *The Theory of Sound* and winner of the 1904 Nobel Prize in Physics.

The previous summer, in August 1912, 670 mathematicians from 27 countries in Europe, Asia, and North and South America gathered in Cambridge for a quadrennial meeting of the International Congress of Mathematicians. The president of the Cambridge Philosophical Society, Sir G. H. Darwin, son of Charles and famous in his own right for his studies of tides, lauded the Cambridge tradition in his opening address. Mathematical physics was alive and well in 1912 and looking toward the future. That very year, the discovery that crystals diffract x-rays had proved that x-rays are waves, like light, and supported the theory that atoms in crystals form orderly patterns. A new device, the modern seismometer, was improving earthquake detection and opening a window to the earth's interior. Air travel brought urgency to mathematical problems in aerodynamics. And everyone was talking about Einstein.

At the same time, the spirit of pure mathematics—math for math's sake— was wafting from the Continent. Bertrand Russell exhorted mathematicians to study the logical foundations of their subject. Hardy stressed mortar and bricks, the need for rigorous proofs. George Neville Watson blended the old with the new. Dot was right to fight her way there: Cambridge University, in 1913, was *the* place for young mathematicians.

Young men, that is. Women math students were few and opportunities fewer. Lectures and tutorials were given at Girton by resident female tutors and by males borrowed from Cambridge and London. The contrast was striking and must have struck Dot. The male tutors were going places. The females at Girton would never go anywhere else.

Women students could also attend lectures in Cambridge proper, but only if the professor agreed. The decision was left entirely to him; him, always him in mathematics; no woman mathematician had ever lectured to Cambridge men. (After the Great War, Dot would be the first.) Russell agreed to let women in, and Hardy, another Honorary Heretic, was an avowed feminist. But some dons refused, while others let them in but ignored them.

Dot took Part I of the Mathematical Tripos at the end of her first year, though she could, and should, have waited till the second. The examination, two three-hour sessions on each of three days, covered pure geometry, algebra and trigonometry, analytical geometry, differential and integral calculus, dynamics, elementary electricity, and optics. She was underprepared, as Surbiton's headmistress Alice Procter had been in her day. Dot's second-class honors was a wake-up call; she doubled down. This may be when G. N. Watson took charge of her mathematical studies. His Trinity Fellowship was ending; he was moving to University College, London. But he continued to tutor at Girton.

If Dot had been a man, Watson might have urged her to change fields. A few years later, another brilliant Cambridge undergraduate, John Desmond Bernal, earned second-class honors on the Mathematical Tripos, Part I. Bernal didn't hear a wake-up call, he heard taps. In mathematics, he might be second class forever, no matter how brilliant he proved himself to be. His tutor persuaded him to switch to the natural sciences. It was a smart career move.[11]

Despite her disappointing showing, it seems no one urged Dot to rethink her ambitions. This may reflect Watson's faith in her abilities, or perhaps he thought it didn't matter much. What prospects did she have anyway?

Girton was founded, in part, to supply teachers for the new, academically serious, girls' secondary schools. But Dot had no interest in a teaching career. Nor did the careers of Girton's resident tutors, Margaret Meyer and Frances Cave-Brown-Cave, inspire her. Meyer, an avid alpinist, kept her hand in mathematics by solving problems posed in *The Educational Times*.[12] The brilliant Cave had scored between the Fourth and Fifth Wranglers in 1898, right after Hardy. This would have earned a man a fellowship, but Girton had no fellowships. Instead, Cave worked in London with, or rather for, the autocratic statistician Karl Pearson, recording barometric readings and teasing out numerical relations between them. After several mind-numbing years, she returned to Girton as a tutor but stayed on cordial terms with her former boss.[13] The turnover in his lab was high; she sent him her best and her brightest.

No woman Dot knew, or knew of, had combined marriage and a career in mathematics. Charlotte Angas Scott tutored at Girton for a few years, then emigrated to America and had a distinguished career at Bryn Mawr. Like Meyer and Cave, she never married. Nor did Phillipa Fawcett, who had scored above the Senior Wrangler; Fawcett became an administrator in the London school system. Another Girton great, Grace Chisholm Young, never held a paying job though she earned a Ph.D. in Germany. She raised six children and coauthored papers with her mathematician husband. Mrs. Ayrton, the famous engineer, built a laboratory in her home.

With nothing to lose, Dot buckled down for the Tripos, Part II. Watson had recently revised and revamped an older colleague's textbook, *A Course in Modern Analysis*. In his gifted hands, "Whittaker and Watson" (as mathematicians still call it) became a clear account of the modern theory of functions and also a succinct course on the special functions of mathematical physics—the Gamma function, Bessel functions, zeta function, hypergeometric functions, and more—that every Wrangler needed to know. Some exercises in the book had been problems on earlier Tripos exams.[14] A century later, "Whittaker and Watson" is still in print.

Dot studied philosophy on her own for its own sake. No career there either: again Miss Jones taught by example. She had been the first, second, and third woman to lecture to the Cambridge University Moral Science Club,[15] and as Girton's mistress she still wrote and lectured on logic despite her administrative responsibilities and fragile health. Yet she was scorned by the philosophers that Dot admired most. Russell went out of his way to avoid this "motherly, prissy, and utterly stupid" woman.[16] G. E. Moore made a withering remark in public that reduced her to tears.[17]

"Dear Sir," Dot wrote to Russell from Surbiton at the end of the summer.[18] Britain had just declared war on Germany, but this gave her no pause. Her letter was all business:

> I hope you will excuse my troubling you, but would you be so good as to tell me to what you refer in your article on "Mysticism and Logic" in the *Hibbert Journal*, p. 793, when you say "logic used in defence of mysticism seems to me faulty as logic and open to technical criticisms which I *have explained elsewhere*" as I should very much like to read your explanation and criticism.... You say there is no intrinsic difference between past and future, but only a difference in their relation to us. But, can there really be an intrinsic difference in anything? All differences in things that we can perceive are surely differences in their relation to us, certainly in material things. It would seem too that the only way things can affect us at all is their relation to us. Please excuse the real liberty I have taken.

The war gave Russell no pause either, but in another sense of the phrase: it jolted him out of his armchair and into the streets. His frenzy was compounded by serious complications in several love affairs. Yet he found time to reply to a Girton student he'd never noticed (alas, his letter is lost).

When Dot returned to Cambridge a month later, much of the male student population had gone off to war, and Trinity's elegant Nevile's Court was a hospital for wounded soldiers. Og kept count of the losses in the *Cambridge Magazine*: instead of its usual 238 students, Clare College had 50, Trinity was mowed in half. And so it went, on and on and on.

To Girton students, cocooned outside of town, the war at first seemed far away. Most were for it. Dot was against it, and Dora was on the fence. But, like the generals of all the warring countries, the students were sure it wouldn't last long. Meanwhile, college life became easier for women students: Cambridge lectures weren't so crowded. In the first year of the war, Dot and a Miss J. Bacon played tennis against Newnham, losing 2–0. Dot rose from secretary to president of the Mathematical Club and continued as secretary of the Music Society and the Society for Little Known Literature. Og invited her to tea and she accepted with pleasure, but backed out at the last minute on orders from Miss Jones. Even so, when Miss Jones retired in 1915, Og was charitable. Although her most recent publication "has not met with the entire approval of the rising school of mathematical logicians whose analytic acumen terrorises the tea-tables of Trinity, [it] has yet won the approval of Dr. Schiller; and attracted no little attention in recent technical discussion," said the *Cambridge Magazine*.[19]

The battlefield slaughter intensified. In *Justice in War Time*, Russell explained why:

> A certain country broke a certain treaty, crossed a certain frontier, committed certain technically unfriendly acts and, therefore, by the rules, it is permissible to kill as many of the soldiers of that country as modern armaments render possible.

Tragedy seeped into Girton's cloistered campus: a brother killed, a cousin missing. Conditions deteriorated. The nightly jugs of milk dried up, replaced by packets of powder. Miss Jones's successor, Katharine Jex-Blake, raised vegetables and pigs on the college grounds and enforced food rationing. The girls "were always hungry, and in winter very cold; we worked huddled around paraffin lamps for warmth and light, and were bitten by mosquitoes."[20] All the while, Dot crammed for the Tripos, Part II.

In the spring of 1916, her college years drawing to a close, Dot weighed her future. Probably at Cave's suggestion, Karl Pearson invited her to work in his lab. The horror! (I shudder at the memory of my summer job in college, transcribing readings from weather balloons over the Grand Cayman Islands.) Any place but Pearson's lab! Dot replied politely on May 12; her post-college plans were uncertain; she would let him know soon (but that second letter, if she wrote one, is lost).[21]

Dot wrote to Bertrand Russell the same day. "In September '14 you were so good as to answer a letter of mine...I hope very much that you will be lecturing next year. May I thank you very much for what you are doing in connection with the war? I have found your *Justice in War Time* particularly helpful."[22]

Dot Wrinch was Girton's only Wrangler in 1916. *The Manchester Guardian* published the names of all of the Wranglers, hers included, but biographies only of the men.

6 DEAR MR. RUSSELL

On June 18, 1916, Dot met her hero at last. "This morning I had a visit from a Girtonian named Miss Wrinch, whom I had never seen before," Bertrand Russell wrote to his mistress, Lady Ottoline Morrell.

> She wants to learn logic but Girton doesn't want her to, and threatens to deprive her of her scholarship if she does. She has just finished her mathematical tripos, and is just the sort of person who ought to do it—she is very keen, but has not a penny beyond her scholarship, and is at the mercy of Miss Jex-Blake, a Churchy old fool. It makes me mad. If she could get £80 a year, she would ignore the College. But she doesn't see how to. If I had it, I would take her on as my secretary. But I don't see how I can manage it.*

Russell wrote letters as if they were e-mails, all day, every day. From his voluminous correspondence, we extract a picture of Dot Wrinch at a crucial point in her life. This particular letter, however, fails a fact-check. Dot had no scholarship for Girton to take away: it had ended with the Tripos. Jex-Blake may or may not have been churchy, but she was a respected classicist and neither old nor a fool. But Dot *was* very keen, *did* want to learn logic, and she hadn't a penny. Russell agreed to teach her. She lined up part-time work to pay him, but Girton came through with a scholarship for her to read for the Moral Sciences Tripos.[1] "The Tripos means doing a large amount of ordinary logic too," she told Og on July 8, but "it would, of course, be very useful, I expect."[2]

Their plan collapsed three days later. Russell was already in trouble with city authorities. On June 5, he had appeared before

---

* The field then called Moral Sciences at Cambridge devolved into economics, philosophy, and psychology.

the lord mayor of London, in the latter's capacity as magistrate, to answer the charges against him. Had he not made statements, in print, "likely to prejudice the recruiting and discipline of His Majesty's Forces"? No, he had not, Russell replied, explaining that refusal to serve and advocating that position were logically distinct. The lord mayor, unpersuaded, invoked DORA, the Defense of the Realm Act, and fined the philosopher £100. Russell's appeal was denied.

The Foreign Office had already canceled his passport, Russell learned later from the Harvard officials who had been expecting him. Now, on July 11, the Council of Trinity College canceled his lectureship. "There was a strong feeling among the senior Fellows that Cambridge was regarded by the world outside as a stronghold of pacifism," his friend G. H. Hardy explained. "This feeling was no doubt exaggerated, but it is undeniable that there was something in it, and that the suspicion with which Cambridge was regarded was due primarily to Russell."[3] On September 1, Russell was forbidden by military order to enter "any restricted territory or fortified district." No more walks along the seacoast, not on the beach, not on the cliffs.

Russell had been living in Trinity quarters. He packed up and moved back to London. His own flat on Bury Street was occupied by a scorned lover, so he moved in with his headstrong, eccentric, older brother Frank (formally, Francis, the Second Earl Russell). Frank lived with seven dogs and his new wife, Elizabeth von Arnim, at 57 Gordon Square, in the heart of Bloomsbury. Rebellious and unconventional, he had been kicked out of Oxford; subsequent adventures included waiting in line outdoors all night for England's first numbered license plate, A 1. (Later, as undersecretary of state for transport, he abolished speed limits. They were reinstated after his death.)[4] At the time we are speaking of, 1916, Frank led the Labour Party in the House of Lords. He stood by his brother on matters of war.

Elizabeth, née (in Australia) Mary Annette Beauchamp, had renamed herself for the narrator of her first novel, *Elizabeth and Her German Garden*. With her German husband, Count Henning August von Arnim-Schlagenthin, she had four daughters and a son. After his death she built a palace in the Alps and began a long affair with—who else?—H. G. Wells. She married Frank on the rebound.

Bloomsbury was family territory: spacious Russell Square, the Russell Square tube station with its 177-step spiral staircase, the four-star Hotel Russell with an elegant restaurant designed by and named for the designer of the *Titanic*, and the popular Lord John Russell Pub, named (perhaps) for their grandfather, the first Earl Russell and twice Queen Victoria's prime minister. All that and more: the family had once owned a good swath of the neighborhood. Boldface names of that era, Leonard and Virginia Woolf,

Vanessa and Clive Bell, the Strachey brothers, David Garnett, Roger Fry, were neighbors and family friends. Roger Fry's Omega Workshop was a short walk away. His sister Margery became Dot's lifelong confidant and friend.

Russell kept his word to teach Dot logic. Modeling himself on Abelard, the peripatetic medieval philosopher, he invited four young people to study privately with him in London: Dot; Jean Nicod; Victor Lenzen, then a philosophy student, later a physicist; and Wallace Armstrong, a physician and disabled veteran turned logician.

Russell's aborted visit to Harvard would have been his second. His first, in the spring of 1914, had been unhappy for him and no doubt for his hosts. Except for three students—Lenzen; a brilliant Greek named Raphael Demos; and Tom Eliot, a young poet—Russell didn't like anyone he met. New England, he told Ottoline, is "the most loathsome part of America."[5] Again, he masked consternation with humor. At a garden party, wrote Eliot, "His laughter tinkled among the teacups."[6]

> I heard the beat of centaur's hoofs over the hard turf
> As his dry and passionate talk devoured the afternoon.
> "He is a charming man"—"But after all what did he mean?"—
> "His pointed ears.... He must be unbalanced,"—
> "There was something he said that I might have challenged."

The three students followed him to England, though not at the same time. Lenzen arrived in 1916 on a traveling scholarship and based himself in Cambridge. "Once a week I journeyed to London and joined a group of four students who met with Mr. Russell for study of *Principia Mathematica* in a room on the first floor of Earl Russell's house....The text for these seminars with Russell was Volume I of the *Principia*."[7]

Coincidentally or otherwise (historians disagree), just as and just when Picasso painted helter-skelter cubes, Schoenberg revised musical tones, and Planck discovered the elusive quanta, Bertrand Russell and the philosopher Alfred North Whitehead tried to rebuild mathematics from the bottom up, a logical edifice to stand for all time.

"I can remember Bertrand Russell telling me of a horrible dream." Hardy tells the story in his late-life memoir, *A Mathematician's Apology*.[8]

> He was in the top floor of the University Library, about A.D. 2100.
> A library assistant was going round the shelves carrying an enormous
> bucket, taking down books, glancing at them, restoring them to the

shelves or dumping them into the bucket. At last he came to three large volumes which Russell could recognize as the last surviving copy of *Principia Mathematica*. He took down one of the volumes, turned over a few pages, seemed puzzled for a moment by the curious symbolism, closed the volume, balanced it in his hand and hesitated.

Rest softly, Russell and Hardy. *Principia Mathematica* has immortal life in cyberspace; anyone can download it, for free, in just a few seconds.

But *Principia* is no longer a guide for the logically perplexed. Its symbols are antiquated, its code too dense even for logicians, its ultimate purpose quixotic. Russell and Whitehead slaved more than a decade, ten hours a day including weekends; and did it, in the end, matter? In 2010, logicians and mathematicians from around the world met at McMaster University in Canada, the home of the Bertrand Russell Archives, to celebrate the book's 100th birthday and to ponder this question.

Principia was a book of and for its time, everyone agreed. Its plan was straightforward, Euclid-style. First, choose axioms and definitions and rules of inference. Next, use them to define whole numbers, rational numbers, irrational numbers, and complex numbers. From this, Russell and Whitehead were sure, all the rest—arithmetic, algebra, trigonometry, calculus, and so on—could be logically deduced. It might take a while, but in principle...

That axioms might, ipso facto, not be up to the task never crossed their minds. Twenty years later, the Austrian logician Kurt Gödel showed that their program had been doomed from the start. But in the meantime, *Principia Mathematica* was *the* text in logic, and comprehending it *the* test for logicians and philosophers. In his brother's house in Gordon Square, Russell and his four students pored over and pondered the book's weighty introduction.

"The introduction, which was written by Russell, contains topics of great philosophical interest, such as the theory of descriptions and the theory of types," Lenzen explained. "This introduction was of especial interest since it represented Russell's work, as he told us. Many years after its publication, this introduction was still being discussed by students of logic."

What *precisely* do we mean by 1 + 1 = 2? Is 1 a symbol, a name, or a number? What, precisely, *is* a number? This matters, said Russell and Whitehead, because the essence of philosophical and mathematical truth lies in symbols and rigorously defined logical relations between them.

In some ancient cultures "numbers" were lengths, in others they were weights. But the discovery of zero, negative numbers, and irrational numbers moved the question from the construction site and marketplace to the academy. What is a number, indeed?

George Cantor, in the nineteenth century, declared a "number" to be a set. Three apples, three oranges, three wizards: "three" is a set of elements containing a triple, a set of cardinality three. Russell agreed, but he was uneasy. For "number" to mean "set" the notion of set had to be rigorous and unambiguous. But sets were problematic.

Galileo had spotted trouble in the seventeenth century. Consider, he said, the squared integers: 1, 4, 9, and so on forever. We can also list them this way: 1^2, 2^2, 3^2, etc., matching them with the positive integers, one to one: 1 with 1^2, 2 with 2^2, 3 with 3^2, and so on. But wait! This one-to-one matching means that the set of squared integers has the same number of elements as the set of positive integers altogether! And that is nonsense: the squares are only *some* of the integers! The set of *all* integers is surely larger than the subset of squares, isn't it?

No it's not, said Cantor, dismissing Galileo's conundrum with a stroke of his pen. We say that a set of three apples and a set of three wizards have the same cardinality because we can match their elements one to one, one apple per wizard, one wizard per apple. That's what cardinality *means*, he said. We must live with the logical consequences. If this implies there are "as many" squares as positive integers, so be it! All that was needed was a name for this infinite cardinality. Cantor called it aleph-nought, $\aleph_0$. (He chose the first letter of the Hebrew alphabet because Greek letters were overused, and printers had it on hand.)

In the words of John von Neumann a century later, "Young man, in mathematics you don't understand things. You just get used to them." Russell got used to this and other weird properties of infinite sets. But sets still made him deeply uneasy: he found a paradox lurking in *finite* sets. The elevator description of Russell's paradox goes like this. Suppose a town has a barber who shaves every man in the town who does not shave himself. Then the town has two sets of men, those who shave themselves and those who are shaved by the barber. Every man in the town belongs to one set or the other, clearly and unambiguously. Every man, that is, except the barber himself.

Russell panicked because the paradox struck a fatal blow to his program of deriving all mathematics from logic. Not because it's a brain teaser, but because it implies, by an intricate chain of formal logic, that *every* statement is true, even false statements like "pigs fly." Even false statements like "2 is greater than 3." So much for the foundations of mathematics!

Russell's paradox can't be named away. But there is a way around it: require all barbers to be women. Or, as Russell put it, require that "whatever involves *all* of a collection must not be one of the collection."⁹ And so type theory was born. The online *Stanford Encylopedia of Philosophy* explains:

> Russell's own response to the paradox was his aptly named *theory of types*. Recognizing that self-reference lies at the heart of the paradox, Russell's basic idea is …[to arrange] all sentences into a hierarchy. The lowest level of this hierarchy will consist of sentences about individuals. The next lowest level will consist of sentences about sets of individuals. The next lowest level will consist of sentences about sets of sets of individuals, and so on. It is then possible to refer to all objects for which a given condition holds only if they are all at the same level or of the same "type."

Type theory: that's what Dot, Nicod, Lenzen, and Armstrong studied with Russell at Gordon Square.

In December, Ottoline invited the philosophical fivesome to Garsington Manor, her grand Oxfordshire estate. "We, foreigners, will never take her in, although it seems we must," laments Elizabeth Hardwick in *Seduction and Betrayal*.¹⁰ "Her invitations, her gifts, her houses, her costumes…For years I thought Garsington, Lady Ottoline's house, was a town name, a resort clever people were always going to or making a point of not going to."

Lady Ottoline Violet Anne Cavendish-Bentinck Morell stood six feet tall, with red hair as long as her pedigree. Inspired by a flamboyant seventeenth-century ancestor, the Duchess of Newcastle, she dressed quixotically and exotically. "The whole story of this lady is a romance, and all she do is romantick. Her footmen in velvet coats, and herself in an antique dress, as they say," said Samuel Pepys of the Duchess.¹¹ Ottoline's contemporaries said much the same about her. She is Priscilla Wimbush in Aldous Huxley's *Chrome Yellow* and Hermione Roddice in D. H. Lawrence's *Women in Love*.

But Ottoline and her husband Philip were also dedicated pacificists in this most contentious of wars. They made Garsington a resort for antiwar artists and writers. Their guests stayed a night, a week, a month, or a year (or two or three), slipping in and out of one another's clothes and one another's beds. Some of the young men did alternative military service on the Morrells' farm.

The students noticed no antics. "Mr. Russell lectured to us in the morning," Lenzen recalled. "We took walks together in the country, and enjoyed social activities with the family and guests."¹² "We all had a great time at Oxford," Dot told Og. "Mr. Russell gave us some new stuff of his, developing some of his points in the Lowell lectures. Nicod is gone to France to be examined [for war service] an nth time." She tried to recruit speakers for The Heretics. "Clive Bell was at the Morrells' and I remonstrated with him abt. Heretics, but he is doing work of great! importance (agricultural!) down there and cannot very well get away. Mrs. Stephen [Virginia Woolf's sister-in-law Karin Costelloe] who was also there is still unavailable."[13]

The next summer, 1917, Russell, Nicod, Dot, and Dot's friend Dora took the weekend walk I described in the Prologue. Today Rosalie Johnson, the youngest of Dot's five nieces and nephews, lives near Pembroke Lodge, Russell's childhood home. On one of her aunt's visits in the 1960s, she told me, they strolled the grounds of the mansion together. "Dear Bertie," Aunt Dorothy sighed. "He invited me on a walking tour with him and Jean Nicod. My mother refused to let me go; *absolutely not!* I begged and pleaded, but she was adamant: *That man has a reputation!* She only relented when I said I'd bring Dora along."

After logic with Bertrand Russell, what next? Ph.D.s were not yet offered in England; a year's research training was the standard mathematical apprenticeship. Russell no longer had formal ties with Cambridge, so Hardy agreed to supervise Dot officially, with Russell unofficially in the background. Hardy soon learned what "wrangler" really meant. When Girton offered Dot a stipend of £40, she drew up a table of her expenses, to the pence, for each of the past four years: "tuition fees, university and library fees, board and lodging, cabs into Cambridge and meals there when necessary; Long Vacation term fees; personal expenses including books, travelling, etc; dress etc; doctors and dentist's bills" and urged Hardy to ask for more. He then requested 80.

Not enough, Dot said, and sent him back again. Hardy raised his request to £120 but refused to go higher. Dot was not, in his view, a remarkable genius. He "was a master who knew to an inch which of his work had value and which hadn't," said C. P. Snow.[14] He judged the work of others on the same strict scale. "Miss Wrinch is able and determined, and I think she should justify the grant," Hardy told Girton's headmistress, "but of course I cannot say that she is so good that she simply *must* be allowed to research—and therefore it would be very inadvisable to apply for an *abnormal* grant."[15]

She got the £120 and settled down to work.

To cut her research molars, Hardy and Russell gave Dot several set theory problems to chew on. At the end of the year she wrote a thesis-like paper,

"Transfinite Types." "Miss Wrinch's notes consist in the main of an interesting development of certain ideas suggested by Hausdorff," Russell wrote to Girton. "They deal with the investigation of series constructed by the 'principle of first differences.' There are a number of new results, and the method employed is obviously a fruitful one, giving possibilities of very important theorems…it points the way to a whole field of new research."[16]

Again Dot commuted between Girton and London, bringing books and journals to Russell, checking quotations and references for papers he was writing, preparing the index for his next book. Her worried doctor urged Jex-Blake to relax wartime rationing and give her more meat, fish, and eggs. Even so, she came down with "a flu" in the spring. Its next, more virulent wave, the Spanish Influenza, would kill millions.

Still in 1917, Dot published a paper, her first, defending Russell against criticisms raised in the journal *Mind*.[17] She lectured "On the Nature of Judgment" to the Cambridge University Moral Science Club,[18] and was elected to the Aristotelian Society, London's leading philosophy circle.

Judgment, said Russell, is not about wisdom, decision-making, or ethics. In his view, judgment was a theory of formal logic, a weapon in his battle with monism, the doctrine that "everything is related to everything else in the universe, and therefore the universe is ultimately a single thing—everything is One."[19] Atomism, by contrast, holds that the world is discrete, particular, and independent of the human mind. Philosophers had debated these isms for millennia.

In the early twentieth century, atomism was on the upswing. Brownian motion and x-ray diffraction had convinced most skeptics that atoms were real. Mendel's rediscovered pea experiments suggested atomistic inheritance. Enthusiasts saw atoms everywhere: "epistemological atomism, or the doctrine of units of perception; linguistic atomism, the use of an alphabet; logical atomism, the postulation of unit propositions; biological atomism, the assumption of discrete cells, genes, etc; and various kind of social, economic, and psychological, as well as physical atomism."[20]

Russell taught logical atomism. This, says the *Stanford Encyclopedia of Philosophy*, "amounts to the claim that the world consists of a plurality of independently existing things exhibiting qualities and standing in relations."[21]

The sword to slay monism, said Russell, is a theory of judgment that strips language to its logical atoms and shows there is more than one of them. Consider the judgment "A is to the left of B." It has three components: an object A, an object B, and the relation "to the left of." These components are independent, since for A to stand to the left of B, B must be a separate object. *Die, Monism!*

Russell's theory evolved: he came to see that a judgment needs a judge. "A is to the left of B" because John sees it and says so: the proper statement is "John thinks A is to the left of B." This led Russell to the "multiple relations theory of judgment" he taught in *Principia Mathematica*.

"The problem about judgment, for both Bertie and Dot," says the philosopher Edwin Mares, "was to understand how to distinguish between believing that Desdemona loves Cassio and believing that Cassio loves Desdemona. The distinction matters: one belief would send Othello into a jealous rage while the other might have been relatively harmless." Mares has studied and written about Dot's contributions to the theory of judgment, which extend Russell's. "On Dot's view," he explains, "Othello's belief that Desdemona loves Cassio includes not only Desdemona, Cassio, and the relation of loving, but also a *function* that puts these elements in the right order. The use of a function of this sort prefigures the work of the great logician Alfred Tarski in his creation of the field of model theory, which was later in the century to become one of the main fields of mathematical logic."[22]

Dot won Girton's prestigious Gamble Prize for her paper on transfinite types. The prize was awarded annually to an Old Girl for outstanding research; the subjects rotated in a three-year cycle. Nineteen eighteen was science and mathematics' turn. The prize committee asked Hardy to judge her paper. He had not changed his mind about her talents. "I do not think any of her work as good as the best of Mrs. Young's, by any means," he wrote. Grace Chisholm Young, the stay-at-home mathematician and mother of six, had—at age 50—won the prize for mathematics the last time around. "I turned up Mrs. Y's Prize Essay in the Quarterly Journal. It contains one really very good theorem, which had been published previously, and there is nothing in Miss Wrinch's work anything like so good. But it is good, solid, intelligent stuff; one part has been published, other parts probably will be in time: and I think that to withhold the Prize from it would be to proclaim a standard for the Prize which would, in existing circumstances, be extravagant."[23]

In the course of the next decade, Dot published the other parts of her paper, as Hardy expected she would. They were well-received, but she developed them no further. Set theory was not, in the end, the passionate cause she was looking for. Gradually, she found herself drawn away from the foundations of mathematics toward broader philosophical problems and Cambridge-style mathematical physics. And to the forms of living creatures.

7

THE SUMMATION OF PLEASURES

Rallies, petitions, speeches, letters, leaflets, essays, books, and still the war dragged on. Russell longed for philosophy's lemon groves, longed to think undisturbed. *Mirabile dictu*, His Majesty's Government, incensed by an intemperate leaflet, granted his wish with a six-month jail sentence. In May 1918, Russell entered Brixton Prison in London.

At last: he could get back to work—his real work, his true calling. And he could do it in comfort. Prison walls were no barrier to Britain's class system. Family connections assured him a cell in the First Division, the wing for incarcerated elites. He could bring his own books and furniture, have meals sent in, and spend his time as he chose. He chose a rigid 14-hour daily schedule: four hours of writing philosophy, four of reading philosophy, four of general reading, and two hours of exercise. Brief visits were permitted in alternate weeks (his brother Frank decided from whom).[1]

But Russell's peace of mind didn't last: emotions soon trumped logic and he found himself unaccountably and irrationally consumed with jealousy. His new mistress, Colette—everyone called Lady Constance Malleson by her stage name, Colette O'Neil—was an actress Dot's age. She and her husband, Miles Malleson, shared Russell's views on love and war.[2] Where was she, what was she doing, and with whom was she doing it? The philosopher tortured himself.

Gladys Rinder, secretary of the pacifist No Conscription Fellowship, handled Russell's official correspondence and smuggled letters to and from Frank, Ottoline, and Colette, many with messages to others buried in them.[3] Dot assisted on the philosophical front, running errands for her imprisoned mentor, visiting when she could, sending him reading material, and writing long letters filled with newsy chatter or discourses on philosophical fine points. "If ϕp is an act of judging or hoping, etc, a certain fact such as 'p is true' will entail χ(φp) where χ will vary with φ. Then if φ is *judging*, χ will be *correct*, or if φ is *hoping*, χ might be *is realized*. This might also make it plausible to explain the fact that judge it *will* rain, and hope it *may* rain, and doubt whether it *will* rain, are of slightly different form, by the difference in the φ..."[4] "Dear Bertie," wrote Gladys, "I hope I've got in Miss Wrinch's symbols all right..." "Good heavens what stuff fr. Wrinch," Frank added.

"We have had such a strenuous week-end with those philosophers," Dot wrote to Russell, reporting on a joint session in London of the Aristotelian Society, the British Psychological Society, and the Mind Association.[5] "It has been so amusing seeing all these people, as I had scarcely seen any of them before." The conference defeated a motion to expel members convicted under DORA, not least him. "As I had sat at your feet, everyone confided their private opinion of you to me!...They ALL said such very nice things and were full of enthusiasm about you."

Eighty people had packed the hall for the weekend's main event, a debate among giants of science. D'Arcy Thompson and John Scott Haldane were born in Edinburgh in 1860 one day apart. Now D'Arcy had a magnificent beard and stood "over six feet tall, with the build and carriage of a Viking and with the pride of bearing that comes from good looks known to be possessed."[6] Haldane had a marvelous mustache and "bore himself with the energetic self-confidence he acquired from his upbringing among the Scottish aristocracy."[7] They had been friends all their lives but diverged philosophically.

D'Arcy was a renowned naturalist and a classicist. He had paid a steep price for this breadth. His first-class honors in the Cambridge Natural Science Tripos would, in the normal course of things, have earned him a fellowship, but Trinity made an exception. "He was too much occupied with too many different projects," says Ruth Thompson, his daughter and biographer. "The very fact that he was so versatile and had so many interests was against him, but the disappointment of never becoming a Fellow of his beloved college was one that he never got over."[8] In the relative obscurity of University College, Dundee D'Arcy set up a museum of natural history, translated Aristotle's glossary of fishes, and compiled a scholarly bibliography of protozoa, sponges, coelenterata, and worms. He also assisted on sea expeditions. All the while, he pondered the shapes of shells, horns, tusks, sponges, tissues, eggs, and leaves. In 1917, he gave in to his admirers and published an 809-page "preface" to the study of biological form. *On Growth and Form* was—is—a masterpiece, "beyond comparison the finest work of literature in all the annals of science that have been recorded in the English tongue";[9] "one of the two or three most brilliant and original books in the life sciences which this century has seen or is likely to see."[10] *On Growth and Form* would join *Principia Mathematica* on Dot's short list of bibles, and D'Arcy would be her mentor, supporter, and friend to the end of his life.

Haldane did not go up to Cambridge or Oxford; he studied medicine in Edinburgh alongside another who would put the training to good use, young Arthur Conan Doyle.[11] While D'Arcy sailed the seas in search of exotic organisms, Haldane descended their depths to experience for himself a diver's malady called "the bends." The decompression stages he recommended are still used today. He descended into coal mines too, in pursuit of "black damp." Black damp, he discovered, was carbon monoxide poisoning. Take a canary down there with you, he told the miners; if it falls from its perch, get out of there. To study breathing at high altitudes, Haldane climbed Pike's Peak in Colorado. His recent invention, the gas mask, was saving lives even as he spoke, in that last summer of the Great War.

Haldane had studied in Copenhagen with Christian Bohr, an eminent respiratory physiologist and the father of Niels and Harald. Christian Bohr and Haldane questioned the common wisdom of the physiologists of their day, which held that oxygen crossed into the blood from the lungs by diffusion across the pulmonary membrane. That mechanical model is far too simplistic, they said. On the contrary, their own experiments suggested that the membrane actively secretes oxygen. By 1918 the controversy had raged for 25 years. Haldane and Bohr lost in the end, but science won: the dispute refined techniques, modified theories, and brought a greater understanding of hemoglobin.

Are physical, biological, and psychological categories irreducible? That was the question, then as now. Is life a brew of chemistry and physics which, if stirred in just the right way, pulls itself together, gets up, and walks off? Or is life something more than the sum of its parts? Is there some other, so-far-missing ingredient?

Something's missing, said Haldane, who spoke first. "Fifty years ago many physiological processes which, from a physical and chemical standpoint, are now seen to be extremely complex and obscure, were regarded as quite simple…There is a prevalent idea that the progress of chemistry, and particularly of physical chemistry, has helped toward an explanation of these processes. This is most certainly not the case."[12] He might have quoted Poul Anderson, had Anderson said it yet: "I have yet to see any problem, however complicated, which, looked at in the right way, does not become still more complicated."[13]

"Dr. Haldane throws down a challenge," countered D'Arcy, "for he tells us, in effect that we have mixed up alien concepts, that in applying the 'mechanical' laws of chemistry and physics to living things we have confused our categories."

But "he gives us so slight an inkling for the moment of his own 'biological hypothesis' that I find little to say about it in the way of Yea or Nay. The main question is, Is it required at all?" Why look for more than physics and chemistry and mathematics, when these sciences have so much to tell us?

If you want a working hypothesis, here's one, said D'Arcy. Many scientists regard heredity as a magical force, but "I for my part look forward, in faith and hope, to the ultimate reduction of the phenomena of heredity to much simpler categories, to explanations based on mechanical lines…that the special science which deals with it has at least found, in Mendel, its Kepler, and only waits for its Newton."

The debate was the main attraction, but there were lectures too. Dot gave a brief talk "on the summation of pleasures," her first in London. Can we determine the degree of pleasure we get from eating chocolates at the opera, she asked, if we know the degrees of pleasure that chocolate and opera give us separately? Yes we can, she answered, and produced a formula for computing it. The Aristotelian Society published the text in its *Proceedings*, but she never put it on her ever-growing resume. Perhaps she thought her formula was trivial. It was, but the question isn't. And the problem, as she posed it—to deduce the combined effect of several stimuli in terms of the effects of the stimuli separately—is a thread through all her later work.

The summation of pleasures problem was 130 years old. Jeremy Bentham, the utilitarian philosopher, had proposed a "felicific calculus" for computing the degrees of pleasure caused by particular actions.[14] In his view, the morality of an action is a balance of pleasure and pain:

Intense, long, certain, speedy, fruitful, pure—
Such marks in pleasures and in pains endure.
Such pleasures seek if private be thy end:
If it be public, wide let them extend
Such pains avoid, whichever be thy view:
If pains must come, let them extend to few.[15]

With a proper calculus, one could compute moralities as if they were betting odds. Bentham left the details to posterity. He left his body too. His mummified remains have been, by the terms of his will, on public display at University College since his death in 1832.

If posterity was bemused by Bentham's will, philosophers were puzzled by his calculus. Can degrees of pleasure be measured? Can they be compared, like degrees of heat? Can they be ranked on a numerical scale?

Numerical scales entered science when Santoro Santoro, a professor of medicine at Padua and a friend of Galileo, put a graduated scale on a thermoscope, as thermometers then were called. The change from "scope" to "meter" signals the sea change in Renaissance science, from philosophical speculation to weighing and measuring. Bentham's calculus would be a thermometer for pleasures. But not a thermometer of the ordinary sort. An ordinary thermometer tells us how hot we are, not how hot we feel.

A decade after Bentham died, a German physiologist, H. E. Weber, gave an experimental subject a one-kilogram weight to hold, blindfolded him, and added smaller weights one by one. Tell me when you feel this getting heavier, Weber said. The subject noticed nothing until the combined weights of the smaller pieces approached a kilogram too.

Sensations of heat, weight, brightness, and loudness are less subjective, perhaps, than judgments like intense, long, certain, speedy, fruitful, and pure, but they too vary from individual to individual. Norbert Wiener, an American philosopher-turned-mathematician, had pondered the problem in 1913, when he studied with Russell in Cambridge. (He too attended Russell's lectures on our knowledge of the external world.) Our perceptions are constructs, not sense data, Wiener concluded. "There can be no science of psychophysics without a proper calculus," that is, without "an analysis of the basis on which this series [sensation-intensities] is put into one-one correspondence with the series of 0 and the positive real numbers, in order of magnitude."[16]

"Any discussion of the question of pleasure seems to be beset with difficulties," Dot began her lecture. She had read Wiener's work but chose to dance around the briar patch of how numbers should be assigned to sensations. Let's just say, she said, that pleasures *can* be ranked, somehow. Then what? How, numerically, do pleasures combine? "There is one very interesting point which I hope I may be allowed to touch on without entering into the intricate and pressing problems ordinarily discussed. *Can the pleasure of several experiences together be expressed in all cases in terms of the pleasures of the experiences separately?*"

G. E. Moore, among other philosophers, had studied this question and thrown up his hands.[17] Dot took it on. "Suppose p is an experience of very intense pleasure, and q is a trivial pleasure such as eating chocolates. Then the pleasure of q given p is, perhaps, very much greater than the pleasure of eating chocolates under ordinary circumstances. On the other hand, the pleasure of p is so intense that q cannot change it."

Obviously, the degree of pleasure of p and q experienced together is *not* the sum of their degrees. Nevertheless, we can compute it, she said. For that, we need a new notion: the influence of one pleasure on another (influence could be positive, zero, or negative). She produced a formula that, translated into English, says: the pleasure of p and q together is the sum of their pleasure separately, plus the influence of each on the other. With suitable axioms, yet to be devised, this formula "gives good promise of yielding elegant and interesting results in a calculus of pleasures," she said.

Dot's first (and last) foray in psychology went unremarked at the time, but science ever circles. Googling "pleasure brain scans" while writing this chapter, I learned that neuroscientists have "observed differential recruitment of brain regions depending on whether subjects ate chocolate when they were highly motivated to eat and rated the chocolate as pleasant, or whether they were highly motivated not to eat and rated the chocolate as unpleasant."[18]

Dot's question lies at the heart of current controversies over quantification, such as *U.S. News & World Report*'s numerical ranking of colleges and efforts in some states to judge teachers by "value-added" metrics. It arises in health care policy. "We can measure and compare health states within a single domain (e.g., hypertension) with units specific to it (e.g., mm/Hg)," explains bioethicist Daniel Wikler, "but that doesn't allow us to measure or compare the health of one person who is blind and infertile vs another who is lame and incontinent. Yet much hangs on how we do this. The usual method is to replace the specific health measure with units of preference or wellbeing— akin to pleasure—which serve as common denominators. Still, 'co-morbidity' remains a thicket of difficulties, in part because of the effect of each experience on the other."[19]

The Aristotelians debated each talk. After Dot's, the minutes record: "There took part in the discussion, the Chairman, Mr. Fox-Pitt, Prof. Alexander, Mr. Demos, Mr. Jeffreys, and Prof. Hicks. Miss Wrinch replied."

Harold Jeffreys and Raphael Demos were Dot's good friends, and maybe more; they had come as her guests. Professors Carr (the chairman), Alexander, and Hicks were past presidents of the Society. St. George Fox-Pitt, a scientist, philosopher, electrical engineer, and inventor complete with white hair, beard and mustache, was Bertrand Russell's first cousin.[20]

"We had your cousin—I mean (I suppose) a cousin of yours (Mr.—Fox Pitt is your cousin?)," Dot fumed to her mentor on paper. "He was a little tiresome about my paper on the summation of pleasures and was all the time bringing up obscure points in electricity...I used an electrostatic analogy....Do you know of anything MORE disappointing than putting an idea forward for criticism and getting UTTERLY IRRELEVANT criticism?...Really Mr. Russell, you should NOT have a cousin who says such bad things about MATHEMATICS."

Russell misplaced her outrage. "I hear St. George blasphemed against mathematics," he chuckled to his brother. "I found an extract about another Pitt…and mathematics: 'Mr. Pitt with his wonderful Quickness of Apprehension, with his strong Understanding, with all his literature, and his honest upright Heart, would not have made the figure he does had he not applied himself to Mathematics with the greatest Assiduity.' (Lady Stafford to her son, Dec. 1789). Tell this to Miss Wrinch—it will comfort her." Frank reported back that it did.[21]

Harold Jeffreys had suggested the disputed electrostatic analogy to Dot one afternoon over tea. She dropped it from the published version of her talk, but soon took it up again. Raphael Demos may have suggested the problem of summing pleasures; he himself was devising a calculus of emotions.

The "unshaven Greek," as Russell unkindly described him, was born in Smyrna in 1892 to a very poor family. He paid his way to and supported himself at Harvard by waiting on tables and doing other menial chores. Demos arrived in England on a traveling fellowship in the spring of 1918, a few months before Russell went to jail. Russell tested new ideas on him and Dot. "I see they don't understand the new ideas I am at," he told Ottoline. "It is no wonder as my ideas are still rather vague."[22]

Before going to prison in May, Russell treated Dot to dinner at Frascati's, a posh restaurant on Oxford Street in London.[23] It was a far cry from Girton's High Table, and Dot was dazzled. Russell didn't play the libertine that evening; he took an uncle-ish interest in his protégé's love life. Life isn't all work, he told her. Take a risk, embark on an adventure.

Gladys Rinder updated Russell in June. "Miss Wrinch, who is a great dear" had asked her to tell him

> that she has "taken your advice and embarked on an adventure but it isn't turning out at all successfully." She was longing to talk to you and in desperation confided in me. I don't think your estimate of her (as she details it) shews your usual insight. The capacity was there, it only wanted arousing. It doesn't sound very enjoyable, particularly since the little green god made its presence felt. The other half is a friend of yours too! She's now in the position of the child who said I wants to go and I wants to stay and I are so miserable! Do send her a message, not that she is really unhappy, but she misses you.[24]

"Give my most sincere sympathy to Miss Wrinch," said Russell, adding that he should like to hear more. In her next letter, Dot told him more.

"With regard to my message a few weeks ago—Do you remember a conversation we had at Frascati's one Sunday night? I am very much disappointed and sick of life. You see, it is apparently no fun unless the relation is symmetric—and unfortunately I am not in love! It's merely distressing I find."

Not in love with whom? Sheila Turcon has catalogued Russell's prison letters;[25] all signs point to Raphael Demos. "I do wonder you never guessed, you know him well, he is one of yr. warmest admirers, & they have worked together," wrote Rinder.[26] "At present she is only very sorry for the 'poor fellow' but as they spend many hours daily together, a sudden change on her part won't surprise me. On Sunday morning I discovered her exploring the nature of emotions with Demos, they are very amazing together I find. They were almost in tears because you weren't available."

Gladys tried to cheer Dot up. "Last night A, I, Dorothy (Wrinch) went to Bach's Phoebus & Apollo at Drury Lane....Afterwards we all went to tea at the Attic, Miles sang to us, to Dorothy's accompaniment, & C. was very gay."

Russell asked Colette to befriend Miss Wrinch too—"I feel she might like it."[27] "Acting on your instructions I've seen something of Wrinch and I like her," Colette reported.[28] "I had tea with her and Demos. He seems rather wicked politically but otherwise angelic."

Colette proved attentive and generous. "On Sunday I'm taking Wrinch to a Stage Society Show (Manfred, with Beecham conducting choir and orchestra in Schumann music). Tomorrow I'm seeing Elizabeth; & Wrinch is supposed to be bringing Demos for E's inspection."[29] Two days later Colette continued, "Elizabeth has a scheme to have you & me & Wrinch & Demos all together at Telegraph House. Does it attract you? (It doesn't me.) But it is clear you must go to T.H."[30]

Telegraph House was the country home in South Downs that Frank Russell bought in 1900, more for its isolation and the view than for the four-room cottage it then was. "The house had been one of the Admiralty semaphore stations in the days when this was the method of rapid communication between Portsmouth Dockyard and the Admiralty in London," Frank explained in his autobiography, *My Life and Adventures*. He bought more land and added a tennis court, planted a laurel hedge for privacy, and built a road up the steep hill. Gradually, the cottage became a mansion with a tower. His second wife had been happy at T.H., but she divorced him anyway. Elizabeth hated the place, but she played the gracious hostess. She and her husband kept tabs on Dot for Russell. "Yesterday I discovered her with Demos and many Bibles," wrote Frank, "in which they were searching for St. Paul's statements on the 'natural man,' without any success. They thought it would help them in their study of the nature of emotion. They wanted you so badly; they say they are making great progress."[31] Tell Miss Wrinch to read Jeremiah 27:9, Russell advised. *The heart is deceitful above all things, and desperately wicked: who can know it?* "It is my favourite text—Freud in a nutshell. Tell her also to let me know when she can what is being discovered about Emotion."[32]

Dot, it seems, was fond of T.H. and fond of its owner too. They would have talked about waterworks. Soon after her father joined the Chelsea Company, it merged with the other private companies supplying London's water to form the Metropolitan Water Board. Frank had opposed the Board in the House of Lords on the grounds that it would be a tool of the water companies. He lost that battle, and "the people of London were robbed and exploited, but they thoroughly deserved it for their apathy, and they are now being punished by paying more instead of less for their water supply."[33]

"Midnight. Had a boiled egg with G. and Demos," wrote Colette.[34] "You'll want to know the Telegraph House news. Wrinch is still there. So is Santayana." Two of Frank's Balliol pals, George Santayana and Edgar Jepson, were T.H. regulars.

Santayana retired from Harvard's philosophy department a year before Demos arrived there; they first met at Telegraph House. Edgar Jepson, a novelist, wrote detective stories; the obnoxious lord in *The Loudwater Mystery* seems modeled on his host. Jepson tried but failed to teach Dot to write a graceful essay.

Ottoline too took Dot under her wing. "Miss Wrinch lunched with me—how very nice she is. I asked her to come down here. Is she very poor?" Russell replied, "She has only what her father allows her, which is little and uncertain and dependent on good behaviour."[35]

"Wrinch is off to Garsington this p.m.," Colette reported, "arrayed in all my Sunday-go-to meeting finery; to be exact, the dove-grey dress & small ribbon hat of our first Ashford. She says she wishes 'to become de-Girtonised.'"[36] "I wonder what you will make of Miss Wrinch," Russell mused to Ottoline.[37] "She seems to have blossomed out since I have been here, but I think it would be rather theoretical. I wonder if you will find her so academic that you can't take any interest in her. I *hope* not. I like her *very* much. She has *very* good brains." In a small photograph in the Lady Ottoline Morrell Collection in the National Portrait Gallery, Dot stands in a doorway. Her hair is de-Girtonized too.

The second Earl Russell pushed Sir George Cave, the British home secretary, to set his brother free. "Of course, young Cave will have to do it," Frank assured Bertie, "because he was my fag at Winchester."[38] Cave agreed, but then, in a sign of the crumbing world order, refused.

Frank wanted his brother out of jail but not back in his home in Gordon Square. He blamed him for the operatic tensions.[39] His wife Elizabeth had struck an odd, if tacit, alliance with her brother-in-law: she sung Colette's praises to the unhappy Ottoline, and Bertie took her side in her battles with his brother. He sleeps with all seven dogs, she complained, and he reads Kipling aloud. Later she took revenge on Frank and T.H. in her masterpiece *Vera*. "Do not marry a novelist," Russell would later warn his children.[40]

That is why the greatest logician since Aristotle spent the month of August 1918 plotting living arrangements for family and friends, "producing more plans and instructions than would precede a medium-sized military operation."[41] Colette would rent "The Attic," her apartment on Mecklenburgh Square, to Elizabeth, who would let Dot live there.[42] Colette and Russell would move into his flat on Bury Street, which, at last, was empty.

Dot was delighted with the plan. Though still affiliated with Girton, she had three compelling reasons to live in London. Harold Jeffreys lived there (he worked in the Meteorological Office); proximity would benefit their work on "scientific method." Dot had also begun research for graduate degrees in applied mathematics with John Nicholson at King's College, London. And she had just been hired, her first real job, to teach mathematics at University College, a short walk from The Attic.

University College had never hired a female mathematician before, but Dot's former tutor, G. N. Watson, had resigned to take a professorship in Birmingham. The war was still on and the male applicant pool was empty. Watson recommended Dot as his replacement. If he had hard feelings, he put them aside. "It appears that Watson fell in love with Dorothy and proposed," their mutual friend Andrew Donald Booth told me. "Dorothy did not like him and arranged for him to meet her Father, a tough Mining Engineer [*sic*]. Meanwhile she had told her father to get rid of Watson for her and according to her story her father kicked poor Watson out of the house!"

But how could Dot do research in two fields and still teach full time? Your duties will not be onerous, the department chairman assured her. Just five lectures a week (on plane coordinate geometry, differential and integral calculus, and elliptic functions), plus four exercise classes, and grading papers. That will leave Saturdays free for research! The decision rested with Girton's headmistress. Dot begged her for the chance. "I should, of course be able to have some help from Professor Nicholson for my research just as if I were doing no teaching."[43]

Cave changed his mind again and released Russell early. Russell rushed to Bury Street. Colette was there but not expecting him, nor was her other lover. A few weeks later Elizabeth von Arnim fled Frank and took refuge with her daughter in California. But one small part of Russell's plan did go through: Dot moved into The Attic.

On November 11 crowds gathered in London to celebrate the armistice. In December, women over 30 with a university degree, or property, or a husband with property, voted in a national election for the first time. As bells rang 1919 in, "we had a New Year's Eve party for which Marcel made hot mulled wine with long sticks of cinnamon." Marcel, a one-name Belgian artist, was Dora's lover at the time. "Dot brought with her a Greek philosophy student called Demos. We descended to the street to hear the midnight bells, where we danced, draped in sundry Oriental covers and rugs from the floor."

"I am so very pleased with life," Dot wrote to Russell from seaside Felixstowe in July. "It is lovely here and I am getting to my work again. It's still scientific method but it's *very* interesting. I feel I am getting on. There are many questions I want to ask you about and if you feel at all interest in S.M. [scientific method] now perhaps you will talk this over in Sept. It would be lovely. R. D. goes back to America in August or Sept. [?] be very pleased at your reference to his article in your Arist. paper."[44]

Raphael Demos returned to Harvard's philosophy department. He retired in 1962 as the Alford Professor of Natural Religion, Moral Philosophy, and Civil Polity, renowned for his lectures and books on Plato. Despite their evidently amicable parting, there is no trace of Raphael Demos in Dot's papers and, his family tells me, he never told them about her.

8   SCIENTIFIC METHOD

September 1919 *was* lovely. Russell and John Littlewood, a Trinity mathematician, had rented a farmhouse in the Dorset village of Lulworth for the last two months of summer. Guests came and went; while in residence, they followed a schedule as strict as Russell's in prison, if pleasanter. Solitary work from breakfast to lunch; afternoons walking in the Dorset hills and swimming in Lulworth cove; dinner cooked by the farmer's wife; and evenings of reading aloud.[1]

Bright star, would I were stedfast as thou art... The next summer, Thomas Hardy would immortalize Keat's stop at Lulworth cove on his fated trip to Rome:

You see that man?—Why yes; I told you; yes:
Of an idling town-sort; thin; hair brown in hue;
And as the evening light scants less and less
He looks up at a star, as many do.[2]

Russell and Littlewood looked up at the stars too, but with less romantic eyes. Was starlight really steadfast? Or was it bent, ever so slightly, by the gravitational force of the sun? And if it was bent, by how much? Newton had predicted a deflection of less than one degree, .875 seconds of arc precisely. Einstein predicted twice that. The bending, if any, could not be seen under ordinary conditions but it could be observed in a total eclipse of the sun. One had occurred recently, on May 29. Frank Dyson, the Astronomer Royal, had dispatched teams of astronomers with cameras to the coast of equatorial west Africa and to northeastern Brazil, the two locations where the full eclipse would be visible. Three months later, word still hadn't reached Lulworth; news didn't travel at the speed of light yet. Russell and Littlewood and their guests waited impatiently. "I wrote to Eddington [from Lulworth] imploring informa-tion as soon as possible," said Littlewood. "Russell remembers a telegram; I remember only a very non-committal letter."[3]

Dot had worked with Harold Jeffreys all summer. They were well matched. Harold was three years older, the only child of village schoolteachers. He studied mathematics at St. John's College, Cambridge, and was a Wrangler in 1913. The next year he was elected a Fellow of St. John's and remained one all his life. Like Dot, he was drawn to applied mathematics, first to astron-omy, then to geophysics, the earth. Like Dot, he was drawn to the philoso-phy of scientific method.

"Scientific method," says the *Oxford English Dictionary*, is a noun phrase meaning "a method of procedure that has characterized natural science since the 17th century, consisting in systematic observation, measurement, and experiment, and the formulation, testing, and modification of hypotheses." Formulation, testing, and modification of hypotheses were the problems they pondered. *How*, from "brute facts"—systematic observation, measurement, and experiment—does a science become a web of interlocking theories? And, since more than one theory can fit any set of facts, how do scientists choose among them?

Sciences evolve from facts to theories, Dot and Harold said, and logic weaves the web. Geometry shows how. The word "geometry" blends the Greek words for earth and measurement; the first geometers must have been surveyors. Geometry became a rudimentary science as measurements piled up and astute observers noted the same facts repeating again and again. For instance, whenever the distances from A to B and from B to C were equal, the angles BAC and BCA were equal too! Gradually, gradually, the notion of *isosceles triangle* emerged in the mind of man. Geometry became a mature science when Euclid wrote his 13 books. Isosceles triangles found their place in a web of pure logic, a consequence of rigorous deductions. Thirteen books, each building inexorably on the ones before it, austere and abstract, not a measurement among them, and few pictures to aid the eye.

"The cowboys have a way of trussing up a steer or a pugnacious bronco which fixes the brute so that it can neither move nor think. This is the hog-tie, and it is what Euclid did to geometry,"[4] said Eric Temple Bell, a twentieth-century mathematician who wrote both fiction and mathematics history, sometimes artfully blending them. But Dot and Harold argued that proper trussing—the logical ordering of cause and effect—is the very essence of science. Relativity theory was a contemporary case in point, subsuming Newtonian mechanics and other predecessors.

To see this at work today, peruse "Is Gravity Real? A Scientist Takes On Newton," in the *New York Times*, July 12, 2010.[5] The featured scientist does not doubt that gravity exists: apples do fall from the tree. But, he asks, what *really* pulls them down? Is "gravity" a fundamental force, a primary cause? Or is it an effect of still more fundamental forces? An effect, he argues. Gravity is "a consequence of the venerable laws of thermodynamics, which describe the behavior of heat and gases." Its place in physics is subsidiary, like Newton's.

Online readers promptly weighed in. "This is all great fun for mathematicians, or philosophers," wrote GR in Berkeley, "but where is the physics? What is needed is at least one single prediction that differentiates it from earlier theories. It doesn't have to be profound, or earth or universe shaking, just unique. Otherwise, pardon this traditional empiricist from asking 'who cares?'" Luke in Waterloo replied, "Now people may question where is the experimental observation? It takes time. If you work in the field of theoretical physics these days, calculations are very complicated and there is still a lot to be done. One could ask Einstein the same thing in 1905."

In September 1919, Dot pondered Einstein *versus* Newton. Which had designed the better truss, logically speaking?

Ludwig Wittgenstein did not come to Lulworth, but he sent a dense manuscript, in dense German. Could Russell help him get it published? Jean Nicod, Dot, and Russell pored over the difficult text. They didn't understand it. Which showed, said Russell, that it is profound.

"Our new party is going to be difficult to manage," Russell wrote to Colette, who was in and out of Lulworth all summer. Nicod had brought his wife along. "Madame Nicod has bobbed hair, is ugly, spirited, advanced, I think rather masterful and inclined to jealousy. Nicod is just the same as ever; so is Wrinch, worse luck."[6] (Colette would have known what he meant by this; I don't.) A few days later, Russell refined his impressions. "Madame Nicod is not really ugly—she is intelligent, intense, Bolshevik, intolerant, without humour or poetry, heartless, passionate and unsatisfied; very jealous of Miss Wrinch, because Nicod and Miss W. talk work together. I dislike her and can't think of anything to say to her. But nevertheless we all get on quite well together."[7]

Dora, one of the new party too, found Thérèse Nicod, with her "fair hair and gray eyes and French elegance, delightful."[8] Dora had not come to Lulworth to talk about relativity, she'd come to begin an affair. She had met Russell for the second time a few months before, when Dot asked her to bring him a small folding table from Cambridge. The conversation begun in the candlelit inn picked up where they left off, but their views had changed in the meantime. Russell wanted a child now—a legitimate child to bear the family name. Dora still wanted children but had ruled marriage out. She would live life on her own terms.

The burdens of aristocracy weighed heavily on Russell's slender shoulders. Frank was childless and intended to remain so, and Bertie's long-estranged wife, Alys, could not bear children. He decided to divorce her and marry a fertile replacement. But whom? Colette didn't want children. Dora did but insisted that *if* she married anyone, it wouldn't be *him*.

Meanwhile, both Colette and Dora were happy to have affairs with the philosopher. At Lulworth, he broke his own rigid rules. "Bertie and I spent much time apart from the company," Dora wrote in her autobiography. "He would hire a boat, from which we would, at sufficient distance from the shore, dive naked; we walked the cliffs talking endlessly. I reproached him for not spending more time with his other guests; he agreed that I was completely in the right about this, but he did not propose to alter his ways."

Nevertheless, in his spare time, Russell wrote passionate letters to Colette. (And to Dora, when Colette was there.) His ambivalence persisted until he married Dora two years later. Colette bowed out for a while, then returned intermittently. "I can remember Colette from occasional visits she made to us during the years at Beacon Hill School," said Dora and Bertie's daughter Kate.[9] "She wore glamorous trailing clothes, strings and strings of long beads and quantities of perfume, and she always brought exotic gifts."

Dot was a friend of and fond of both women. Did their rivalry pain her, or did it amuse her? It has been suggested that Dot wanted Russell for herself,[10] but the evidence is insubstantial. Perhaps, buried in work, she didn't much care. And, it seems, she had lovers of her own.

"I did enjoy myself at [Lulworth] most frightfully!" Dot wrote to Russell from Surbiton afterward. (She left The Attic in January, when its roof caved in.) "Thank you so very much for all the lovely times we had. You really are the most delightful person in the world to stay with! September has been one long pleasure and I shall never forget it."[11]

"There were three possibilities," Arthur Stanley Eddington, the leader of the Africa expedition to view the eclipse, told the Royal Astronomical Society in December.[12] Dot would have attended that meeting, at which she was nominated for election to Fellow in March. Harold Jeffreys and John Nicholson were Fellows already.

"There might be no deflection at all; that is to say, light might not be subject to gravitation," Eddington explained. Or "there might be a 'half-deflection,' signifying that light was subject to gravitation, as Newton had suggested, and obeyed the simple Newtonian law. Or there might be a 'full-deflection,' confirming Einstein's instead of Newton's law." He paused for effect, then continued dramatically. "Three days after the eclipse, as the last of the calculations were reached, I knew that Einstein's theory had stood the test and the new outlook of scientific thought must prevail."

The meeting was contentious. Several astronomers objected that the photographs were too fuzzy to measure the difference between these very small angles. Others stressed a point of logic. "A implies B" and "B implies A" are not the same: Einstein's theory predicted a shift of 1.75 arcseconds, but a shift of 1.75 arcseconds did not prove Einstein's theory. There might, they insisted, be a better, simpler explanation that no one had thought of. Still others resisted the change of worldview. Sir Oliver Lodge refused to swallow (his word) the new views of time, space, and force. "That it should be any assistance to replace a straight line by a crazy geodesic is surely puzzling," he objected. Amen, said Ludwig Silberstein. "To say the least, it seems premature to discard every possibility of non-gravitational refraction....Einstein introduces at the very beginning the crudest unanalyzed concepts of rigid bodies (as measuring-rods) and ordinary clocks. A theory erected on such bases cannot, manifestly, claim the capacity of reforming our usual concepts of time and space, or even only of shedding some usefully critical light upon them."

How do scientists choose among competing theories? Experiments, or observations, may not give clear results. Even crystal-clear results can't prove a theory; at best they can support it. Eddington told his audience that the evidence had persuaded him, but that wasn't true. He, the leader of the Africa expedition, had thought the trip a waste of time and money. Clear or fuzzy, the photographs were irrelevant. He knew beforehand that Einstein was right.[13]

Eddington agreed to lead the trip, not as a giant step for mankind but to avoid the wartime draft. He wouldn't have served in any case: a devout Quaker, he had refused on grounds of conscience. But this principled stance alarmed his friends and colleagues. Conscientious objection would save the young genius for science, but with a stigma. Pacifists were untouchables in fervid wartime Britain; just look at Bertrand Russell! Eddington's friend Dyson, the Astronomer Royal, persuaded His Majesty's Government to exempt the brilliant young astrophysicist from the draft in the national interest, to lead an eclipse-viewing team.

But why was Eddington sure Einstein was right and Newton wrong, without looking at such evidence as could be found? John Keats had already explained this in one succinct, immortal phrase: *beauty is truth, truth beauty*. Einstein's theory was more beautiful.

But is truth beauty? And if it is, why? Dot and Harold delved into the puzzle. Faced with choices, scientists select the theory they think is most likely true. Often, this is also the most beautiful theory (though just as chance favors the prepared mind, beauty in science favors the trained eye). The most beautiful theory is often the simplest. Should Keats have written *simplicity is truth, truth simplicity* instead?

What's going on here? they asked in effect. "Is the prevalence of these simple and accurate laws due to the nature of our investigation, or to some widespread quality in the external world itself?" It is true that simple laws are easy to work with, but that's not the only reason scientists choose them. If you look at how scientists actually choose between theories, you find they assume that the simple laws are very probably right. In their words, "scientific practice seems to require the assumption that an inference drawn from a simple scientific law may have a very high probability, often not far from unity."

Scientific method—how to choose one theory over another—is not an art, Dot and Harold said, it's a science: the mathematical science of probability.

For the Moral Sciences Tripos, Dot had studied probability theory with W. E. Johnson, a logician at King's College, Cambridge. Though he published little, his influence was vast. A widower, Johnson lived in King's, where he was a Fellow. Ill health and shyness kept him near the fire (and the piano), always wrapped in the same red shawl. He had a "lovable personality," and those who attended his lectures "were infected with his exacting subordination of originality to clarity and truth."[14] His circle around the fire included (not simultaneously) many now famous: the economist John Maynard Keynes, the philosopher Charles Broad, the mathematician Frank Ramsay, and Dot Wrinch. Her summation of pleasures formula drew on notions and notations she learned at his feet.

In 1916, "probability" meant different things to physicists, gamblers, geneticists, astronomers, astrologists, philosophers, and the proverbial man in the street. Even Russell, a stickler for precise, exact definitions, used "probable" loosely, as a synonym for the vague term "credible." The first task of the logician, surely, was to define "probability" rigorously. But that project was floundering. Could a single definition meet the demands of all these disciplines? Some said no. Others said yes, but then clashed on what it should be. The debate boiled down to two approaches: "Do we try to determine statistical chances of obtaining certain samples from some real or hypothetical population, or do we try to set up a mathematics of degrees of belief?"[15]

Gamblers and Mendelian geneticists argued for samples, astronomers and physicists for estimates of prior likelihood. (If Dot had worked in Karl

Pearson's laboratory after Girton, she might have landed in the thick of this debate a few years earlier. Pearson, editor of the journal *Biometrika*, battled the Mendelians in and out of its pages.)

What is the probability of heads in a flip of a coin? Most coins are biased to some degree; best to do a "frequency" test, flipping the same coin again and again, and record the results. If you get 600 heads, say, in 1,000 flips, you would expect heads on the next throw with a likelihood of 60 percent.

But astronomers and physicists asked different questions, questions like "what is the probability that the earth's core is molten?" To answer it, you don't do a series of identical tests, you base your answer on what you already know about the earth's composition.

Use what you know? That's subjective, said the "frequentists." You know many facts and you weigh them according to *your opinion* of their relative importance. Science must be objective. You shouldn't base theory choice on your personal assessment of the odds.

Before Dot and Harold, explains historian David Howie, that's where writers on probability theory got stuck.[16] In their first, 1919, paper, Dot and Harold made a breakthrough, along the lines of her argument in the summation of pleasures: *don't go there.* Assume "prior" probabilities can be assigned somehow. Ask not what probability *is*, they said. And don't ask what it means. Line up the probabilities on a number scale, a scale with a beginning and an end, say 0 (0 percent for impossibility) and 1 (100 percent for absolute certainty).[17] Obey these three axioms:

- every probability corresponds to a number;
- the larger the probability, the larger the number;
- if two propositions are mutually exclusive, then the probability that one or the other is true is the sum of the probabilities of each. (The probability that either it will rain tomorrow or it won't is 100 percent.)

Dot and Harold, says Howie, by "starting from axioms concerning comparability rather than the combination of propositions, were able to...widen the range of probabilistic induction to the predominant situations in which equally probable sets of alternatives could not be constructed." Within this framework, "inverse probability"—a mathematics of degrees of belief—long out of favor, could be discussed dispassionately.[18]

Dot and Harold made probability the cornerstone of their scientific method. Over the next few years, they elevated the apparent parallel between simple scientific laws and probable ones to a "simplicity postulate" and used it to expand their scientific method to a broader theory of knowledge. And they applied their ideas of scientific inference to seismology.

Eight thousand earthquakes a day kept the seismographs scribbling, but what were they saying? The modern seismograph was only 10 years old; interpretation was in its infancy. Professor Herbert Hall Turner at Oxford, publisher of the *International Seismological Survey*, was swamped with undeciphered data from far-flung research stations.

Harold hoped seismology would reveal the structure of the earth. In Jules Verne's tale *A Journey to the Center of the Earth*, the fictional Professor Lidenbrock and his nephew Axel travel down, down, down through layers of increasingly ancient rocks, meeting increasingly prehistoric creatures, to the earth's hollow center. Hollow—is the earth's center hollow? No one believed that even when Verne wrote it in 1864. But if it isn't hollow, what is it? Is the earth solid through and through? Or is it a crust with a molten core? If the earth has a crust, how thick is it? Is it homogeneous, or is it layered? What is it made of? One could answer these questions, Dot and Harold surmised, by analyzing seismic waves from several stations. Dot volunteered to come to Oxford to help Professor Turner from time to time; she arranged to lunch in Somerville College.[19]

Her plans were upended by a tragedy in faraway Oppau, Germany. Early in the morning of September 21, 1921, 4,500 metric tons of ammonium sulfate and ammonium nitrate stored in a silo of the Badishe Anilin und Sodafabrik spontaneously exploded. The blast, heard in Munich 300 kilometers from Oppau, held the record for man-made disasters until Hiroshima. Hundreds of people were killed and thousands wounded.[20] Dot informed Jex-Blake that same day that she would do her seismological work at Girton, not Oxford. With Harold, not Turner.

Analyzing the seismographs produced by the explosion, Dot and Harold found that a layered crust, with one layer granitic, would match the data. That was not all they discovered. Seismologists knew that some seismic waves travel through the earth (in two perpendicular directions), but others roll over its surface. Lord Rayleigh, the author of *The Theory of Sound*, had identified the surface wave now named for him. In 1911, A. E. H. Love, an Oxford mathematician, proved—mathematically[21]—that if the earth has a crust and the crust is layered, there must be a second type of surface wave. Dot and Harold showed that he was right. Professor Love has shown, they wrote, "in an important section that has hitherto received too little attention, that there is a type of purely distortional wave that can travel only along the surface in such a crust...it will be convenient to refer to this as the Love wave."[22] Love waves, as they are still called, do most of the damage in earthquakes.

A slim volume, *Lectures on the Principle of Symmetry and Its Applications in All Natural Sciences*,[23] brought Dot an epiphany. She recorded the date, March 31, 1921, on the title page. The author, F. M. Jaeger, a Dutch chemist, could have taught my symmetry course: kaleidoscopes and repeating patterns, group theory, crystallography, symmetry everywhere. Under the date Dot wrote "p 3." I didn't notice this when she gave me the book 40 years ago; now I turn to the page. I see what struck her so forcibly: symmetry is the combined effect of several stimuli together, said Jaeger, a summation of visual pleasures:

> The splendour and fascinating beauty of a great number of living creatures: *radiolaries, medusae, diatomeae, corals, starfishes,* of innumerable *flowers*; that of the splendid forms of many *crystals* and of the figure produced in Lissajous' wellknown experiments with *combined harmonic vibrations,* in *vibrating membranes* (eidophone) or *metalplates* (Chladni's *sand-figures*)...are in each case caused by the action of symmetrical repetition.

An epiphany, but in years to come symmetry's causes would fade into the background of her fertile mind, and only its beauty would remain.

There are two kinds of analogy, Dot told the Aristotelian Society later that spring. *Suggestive Analogy* is metaphor, a jumping-off point. *True Analogy* looks behind appearances. The electrostatic example that had so riled Fox-Pitt was a case in point; she revived it. "If a conductor consisting of a sharp straight edge be placed in any field of force what is the modified field of force in its immediate neighborhood?" This simple situation crops up everywhere in disguise, she pointed out. "This is the problem before the architect designing a lightning conductor on the flat roof of a house, and the solution is worked out in mathematics with precision....Now in thermodynamics the architect or the botanist may wish to know the distribution of temperature, say in a large gallery or greenhouse with hot water pipes along the side and a special heating apparatus at some point along the wall. If the physical conditions are suitably adjusted, one and the same solution will serve for either problem."

Harold Jeffreys and John Nicholson were in the audience this time. So was St. George Fox-Pitt. Again he "took part in the discussion," but we don't know what he said. Again Russell wasn't there. He wasn't in jail this time, he was in China.

Photographs by Harold Jeffreys. By permission of the Master and Fellows of St. John's College, Cambridge.

Peking University had invited Russell to lecture, and Dora agreed to go with him. In August 1920, after secretive plotting to which only Og and Dot were witting,[24] they left for China as a couple. They agreed to marry when, and only when, an heir was on its way.

Russell left Wittgenstein's manuscript in Dot's hands. "Wrinch became a key figure," says historian of mathematics Ivor Grattan-Guinness.[25] "After it was rejected by the Cambridge University Press and by a publisher and some journals in Germany, she placed it with the *Annalen der Naturphilosophie*, edited by the chemist and zoologist Wilhelm Ostwald—an improbable venue, secured by Ostwald's high opinion of Russell and the promise of an introduction from him." (Og published an English edition two years later, with the title *Tractatus Logico-Philosophicus*.)

Dot and Harold were constant companions, writing papers together, singing in the Philharmonic Chorus, and planning—with a small group of Cambridge scientists—the postwar organization that became the National Union of Scientific Workers.[26] Dot's name appears next to Harold's in a guest list for a reception for Einstein.[27] Harold grew close to the Wrinch family; his photographs are the only ones we have of them in those years.

Not surprisingly, some observers thought they were engaged.

But Dot also went about with her dissertation adviser, John Nicholson. "It's lovely being here, real big mountains," she scrawled on a postcard in September 1920. The card reached Russell in Beijing. "I hope you (Sie, not Du) are flourishing. Christiania and a glimpse of the International Council of Women, Stockholm to see the mathematicians, Copenhagen to see Bohr family, Amsterdam and then [illegible]: a great programme! All our hearts, DMW, JWN."[28]

Dot was also, Colette wrote to Russell in China, being seen about with his brother Frank.[29] Russell must have raised this point in a letter, as Frank replied, "I have not seen the elusive little Wrinch again, though she seems to spend as much time in London as at Girton. I did not know a don had so much freedom of movement in term time."[30]

Dot will—I jump ahead—marry John in 1922.

"I suppose his jilting was what brought him into Jones's hands," a friend of Harold's wrote in 1925, after a chance encounter.[31] Ernest Jones was a psychiatrist, Freud's leading British disciple.

Dot gave talks and wrote papers on scientific method through the 1920s. Sir Harold—he was knighted in 1953—pursued the mathematical questions. He was the leading voice for Bayesian statistics in the twentieth century and the author of landmark books. In them and elsewhere he always acknowledged Dot's influence. "I should like to put on record my appreciation of the substantial contribution she made to this work, which is the basis of all my later work on scientific inference."[32]

III BIOLOGY IN TRANSITION

While I have sought to shew the naturalist how a few mathematical concepts and dynamical principles may help and guide him, I have tried to shew the mathematician a field for his labour—a field which few have entered and no man has explored.

—D'ARCY THOMPSON, *On Growth and Form*

9 THE SPICULES OF SPONGES

Morphology: that was the question. How the Whale got his throat, how the Leopard got his stripes, how the Camel got his hump. But Darwin, not Kipling, held sway in turn-of-the-twentieth-century Britain: Natural Selection surely fashioned each and every feature of each and every creature.

Zoology "was still almost wholly occupied with problems of phylogeny and comparative anatomy, that is, with the apportioning out of evolutionary priorities and the unraveling of relationships of descent," Sir Peter Medawar said of D'Arcy Thompson's time.[1] Most zoologists "had no real curiosity beyond the evolutionary pedigree of an organism or an organ: any inquiry into the action of contemporary physical causes seemed to them to belong to some other science."

Was natural selection really such a totalitarian, omnipotent, micromanaging engineer? D'Arcy had doubted all along. "Soon after leaving Cambridge, when he questioned some pronouncement of Darwin's in a lecture," says his daughter Ruth, "an old don took him aside and said: 'D'Arcy, you may think these things, but you must not say them. It is not the time, and what is more, it is not the way to get on.'" At last, in 1917, he said them.

Consider sponges, those strange animals tethered like plants to the ocean floor. There are 5,000 species, most of them microscopic. Their skeletons are meshes of spikes, called spicules, of wildly different shapes and textures. In some species, the spicules are long, in others short; some are thick, some thin. Some are straight, others curved or spiral. Some end in teeth, like pitchforks. Why is that? Each and every twist, turn, and forking served, or once served, some adaptive purpose, selectionist fundamentalists replied. (The claim is heard today on the outer fringes of evolutionary psychology.[2])

D'Arcy examined a species of sponge with rigid triangular spicules; the triangles have curved sides and elongated corners. Their

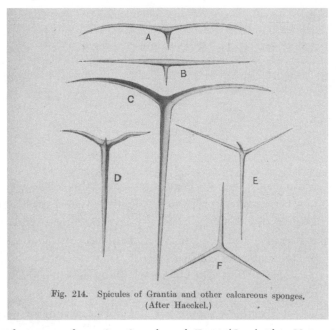

Fig. 214. Spicules of Grantia and other calcareous sponges.
(After Haeckel.)

Spicules of sponges, from *On Growth and Form* (Cambridge University Press, 1917).

composition is calcarous, crystalline. They take this shape for an elementary reason, he surmised. The cells of the sponge don't form tissues, they pack together like pennies. The crystalline spicules grow in the gaps between the cells and fill them. When the cells die they disintegrate and leave the triangles behind. Natural selection had nothing to do with it. (See figure 3.)

"I shall never forget his description of the triradiate spicules of the calcareous sponges with their rays forming at coequal angles of 120 degrees," a student told Ruth, "because for the first time I became aware that mathematics may be applied to give precision to biological observations and thus to open up a fascinating vista of speculations."

Which brings us to the papers Dot presented to the University of London for her graduate degrees.[3] She didn't mention sponges; the papers are pure mathematics. But sponges had shaped her work just as they shaped their spicules. Shaped, and vanished.

Dot's thesis subject was "asymptotic expansions," a subfield of infinite series. Her adviser, John Nicholson, put her to work on them because scientists needed them. That is, he needed them—he and Arthur Dendy, his biologist colleague at King's. Dendy was the world's leading spongeologist and the author of the authoritative *Studies on the Comparative Anatomy*

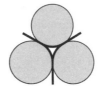

| Three sponge cells. | The spicule grows in the space between them and is shaped by it. | The cells die; the spicule remains. |

Figure 3 The stages of spicule formation, suggested by D'Arcy Thompson.

of Sponges. His popular textbook *Outline of Evolutionary Biology* had been reprinted again and again. But, like D'Arcy, Dendy questioned natural selection as evolution's sole mover. "What can it matter to the sponge whether the number of the teeth be three or more or less, or whether the teeth at the two ends of the spicule be equal or unequal?"[4]

The 1905 Sealark expedition in the Indian Ocean had dredged up a new microscopic sponge genus, the *Latrunculia.* Its spicules were knobby. Why knobs, and why did the knobs appear on the spicules just where they did? Dendy wondered. To say the knobs are "the result of adaptation to the requirements of the sponge as a whole, produced by the action of natural selection upon variation in every direction" begged the question.[5]

Professor Dendy knew his classics. Aristotle had sorted sponges into three types, according to their texture.[6] Loose and porous sponges comprised one class (the hardest and largest); close-textured sponges (the softest) comprised another; and "the sponge of Achilles," used to line helmets "for the purpose of deadening the sound of the blow," the third. Nurture—the environment— accounted for these differences, said Aristotle. "Sponges found in deep calm waters are the softest, for usually windy and stormy weather has a tendency to harden them and to arrest their growth. And this accounts for the fact that the sponges found in the Hellespont are rough and close-textured."

A spicule vibrates in the sea like a string vibrates in the air, Dendy reasoned. A string vibrates around stationary nodes whose positions can be calculated with Lord Rayleigh's mathematical-physical theory of sound. Spicular knobs looked liked stationary nodes to him. If this analogy was True, then Rayleigh's theory, adapted from sound to water, should predict their locations. Dendy enlisted John Nicholson. He, Dendy, would measure the actual positions of the knobs on the Sealark samples, and Nicholson would calculate their theoretical positions. If measurement matched theory, they'd have a case that the ocean was the architect. As a boy, Nicholson had collected butterflies and other insects; Dendy's project must have appealed to him.

With a *camera lucida*, Dendy enlarged microscope images of *Latrunculia* and reflected them onto a sheet of paper. Then he traced the reflections and measured the positions of the knobs. Meanwhile, John figured out, theoretically, where the knobs should appear. "On the whole, it is somewhat surprising to find how close the agreement between the observed and calculated position of the whorls really is," they reported in understated tones. "We have here a new factor in ontogeny."[7]

But there's a gap in the paragraph above. Between figuring out, theoretically, where the knobs should appear and matching the numbers lies a mountain of computation. John had found general formulas for the locations of the knobs on the spicules. But extracting the exact locations—precise numbers—from the formulas, was no easy task. He did it for this special case, but the general problem remained. He gave it to Dot.

If, for some reason, you want to know the sine of 67 degrees, you enter 67 in your calculator and press a button. You get the answer to any number of decimal places in a wink. Your calculator finds it by executing a built-in algorithm, for example adding the first several terms of the infinite series for the sine function,

$$\sin(x) = x - x^3/3! + x^5/5! - x^7/7! + \ldots$$

where 3! (three factorial) stands for 1x2x3, 5! (five factorial) for 1x2x3x4x5, and so on. The more terms the calculator sums, the greater its accuracy, because the series converges to the sine function itself.

The functions John used to solve the spicule problem are much more complicated than $\sin(x)$, and no infinite series converge to them. But in some cases there are series—called asymptotic expansions—that come close enough for practical purposes like calculating the locations of spicular knobs. Here the problem splits in two: finding the asymptotic expansions and computing their values. Dot worked on the first of these.

"'Asymptotic expansions,'" John explained in a public lecture at King's,[8] "are the most frequent type of connecting-link between Pure Mathematics and subjects of mathematical physics such as hydrodynamics, special problems in distribution or motion of electricity, the diffraction of light round obstacles, and acoustic vibrations—to mention only few of the main subjects concerned." In other words, asymptotic expansions reveal true analogies.

"We may say," he continued with evident pride in his protégée's accomplishments, "that such an [asymptotic] expansion is an expression which, to a certain degree of accuracy, gives the same numerical values as the function which actually occurs in the problem. It is often very difficult to find, though many valuable ones have been found recently, and are still being found, by quite new methods of considerable generality."

In 1920, Girton awarded Dot its new Yarrow Research Fellowship. For the next four years, she could follow her burgeoning interests wherever they led her. Sir Alfred Yarrow, a wealthy engineer, inventor, shipbuilder, and Girton benefactor, invited her to tea and asked about her work. His interest was genuine. Like the *Latrunculia*, he had been shaped by the sea; he could no longer sleep on dry land. The aging mariner built a swinging cot with a rocking mechanism and took it with him wherever he went.[9]

In 1920, the University of London awarded Dot a master of science with distinction and a doctor of science (D.Sc.) the next year. As examiners for the latter, the university appointed Nicholson and Whittaker ("or, failing him, Watson"), with A. N. Whitehead as "referee." What does that mean, *referee*? I asked Ivor Grattan-Guinness, the British mathematics historian. He asked the UCL archivists; they didn't know either. So we guessed. Women candidates were few, and those seen about with their research supervisors were fewer. Appointing a tie-breaker would fend off suspicions of conflict of interest.

But no referee was needed after all; everyone was impressed with Dot's research. On the strength of that, she became the first woman in Cambridge's 800-year history to teach mathematics to men. Hardy encouraged her to develop her lectures into a monograph: "I understand that it would in all probability be possible to have it published in the Cambridge Mathematical Tract Series," she told Girton in her first fellowship report.[10] "I have various applications of this work on Asymptotic Expansions in progress, in acoustics and in connexion with certain morphological problems in biology."

"Computers" were cranking out the numbers and compiling tables for her. In the 1920s, computers were human beings, hobbyists with a passion for tables and interpolations and complicated arithmetic. The British Committee on Mathematical Tables decided which ones should be computed and supervised the computations. At the Committee's behest, computers calculated, by hand, values of the very complicated functions in "Whittaker and Watson." Watson and Nicholson were members of the Tables committee; Nicholson chaired it from 1920 to 1931. Dot joined in 1921—the only woman member in its 94-year history.[11] Her father, Hugh Wrinch, computed occasionally and they published two papers together.[12]

That year Dot had also, she told Girton, worked with Harold Jeffreys on probability theory and problems of scientific methodology, published a paper on "Relativity and Geometry," submitted another for publication, and was writing a third. She did not mention her D.Sc., or that she had registered at King's for a second doctorate, a D.Lit. (doctor of literature) in scientific methodology. The Board of Studies in Philosophy approved Professor Nicholson's request that King's waive its requirement that a candidate first earn a master's in that field. She was to complete the D.Lit. in 1923.[13] But this project melted away.

John William Nicholson's father was an ironworks clerk, his mother "the daughter of a painter." He was born in Darlington, England, in 1881 and educated there. "As a boy he was very fond of country rambles and was an enthusiastic entomologist," says his biographer, William Wilson.[14] "One suspects that his enthusiasm for anything of a mathematical character deprived the world of a great biologist."

After secondary school, John studied mathematics and physics at Owens College in Manchester (later renamed the University of Manchester) to prepare for the rigors of Cambridge. Arthur Stanley Eddington, the astrophysicist-to-be, enrolled for the same reason at the same time. Together they went up to Trinity in 1902 and sat the Tripos two years later. Eddington won the top prize, Senior Wrangler, the first student in Cambridge history to manage that feat in such a short time. Nicholson came in 12th.

Before Max Planck announced the quantum in 1900, Newton's mechanics had seemed complete; within the decade, physics was upended on all scales, from the atom to the universe. John spent the next several years in the thick of the action, first at the Cavendish Laboratory in Cambridge and then at Queens University in Belfast. He was appointed professor of mathematics at King's College, London, in 1912. "Like other mathematical geniuses," says Wilson, "he was apt to get absorbed in some problem when he ought to have been thinking of the lecture he had to deliver....they [the students] did not mind his lateness, since they got so much from him even during half an hour."

John Nicholson's research at that time concerned the spectra of stars and the sun, and what they revealed about the structure of the atom. In 1911, he presented a model of the atom to a meeting of the British Association for the Advancement of Science; he worked on it for the next three years. "While Nicholson's atomic theory had an inherent interest and significance," says the physicist and historian Russell McCormmach, "especially as the first theory capable of calculating spectra from the internal motions of precisely defined dynamical structures, and as an apparently successful application of the quantum theory, it has become clearer with the perspective of time that the probable point of Nicholson's greatest influence on the course of physics did not lie in 1914, or anytime later, but in the winter of 1912 when Bohr read [Nicholson's] *Monthly Notices* articles on the constitution of the solar corona."[15]

Astrophysicists visiting Drachenfels, Germany, September 1913. Left to right: Frederick Stratton, John W. Nicholson, K. Schwarzschild, Frank Watson Dyson, H. Knox Shaw, and Arthur Stanley Edison.

Emilio Segre Visual Archives, gift of Martin Schwarzschild. Niels Bohr Library and Archives, American Institute of Physics, College Park, MD.

In the summer of 1913, while Dot was packing for Girton, John Nicholson traveled to Bonn with Eddington and Dyson for the Fifth Conference of the International Union for Co-operation in Solar Research. After the meeting, they posed for a photograph at the Drachenfels Castle. Eddington smirks atop the pony at the right, Dyson is third, John fifth. Reserved and unsmiling, he has dark wavy hair and a kind, gentle face. He is 32 years old and strikingly handsome.

John left King's for Balliol College, Oxford, in 1921. (Say Bailey'll, says the college website, "emphasis on the first syllable. Not Bally-all, Ball-oil or

Bailey.") Two years later, Dorothy Sayers's first mystery, *Whose Body?'* made the ivy-covered college a household name, and Lord Peter Wimsey its most famous graduate.

"There could have been no sign of any problems when Nicholson was hired," John Jones, Chemistry Fellow and Fellow-archivist, told me. "They vetted Fellows fully, then as now. He must have been very conventional—well, very much like them."

Dora and Bertie returned from China in the summer of 1921 to find themselves homeless. They had rented a place in London, but three years after the armistice passions still ran high, and the landlord changed his mind: he would not house a traitor. Dot had been renting a weekend cottage in Winchelsea; she lent it to them until they found a home of their own.

Russell's divorce came through a few days after the Oppau explosion. The brief, grudging wedding ceremony was held at the Battersea Registry Office. The heavily pregnant bride wore a black cloak and carried a bouquet of sweet peas wrapped in newspaper. After tea in a nearby cafe, the newlyweds returned to Dot's cottage. They moved to a London townhouse just before their baby's birth in November.

The next summer, the Russells rented a cottage in Porthcurno, Cornwall, and then they bought it. To the end of her life, through thick and thin, Carn Voel would be Dora's true home. Dot "joins me in thanking you and Dora for the really wonderful time you gave us," John Nicholson wrote to Russell after a visit. "Personally I hope it is only the first of a long series (to come) of happy times for the quartetto (or rather quartette)."[16]

Dot and John were married in the Church of St. Mark, in Surbiton, on August 1, 1922, with relatives, friends, the Astronomer Royal, and the mistress of Girton College attending. With a year of her Yarrow Fellowship to go, the bride continued to live at Girton. The groom was tethered to Balliol; they shared the Winchelsea cottage.

AN OXFORD AND CAMBRIDGE MATCH, wrote "a Woman Correspondent" for *The Manchester Guardian.*[17]

> The bride, Dr. Dorothy Wrinch, is one of Girton's most successful women. She is quite young, and is only at the opening of what her friends believe will be a notable career....Dr. Wrinch is a good type of modern university woman, for she has had many interests outside her special work. She is interested in social questions and in politics and is a fine musician.

The London Times noted her expertise in the new science of seismology.

Bertha Swirles, Harold Jeffreys's future wife, was a student at Girton at the time. We girls couldn't see what she saw in him, she recalled, but what do 18-year-olds know?[18] Frank Russell would call him "Old Heavyweight."

"I fear you will think this letter somewhat premature," Lynda Grier wrote to Dot a few weeks after the wedding.[19] Grier was the principal of Lady Margaret Hall, one of Oxford's five women's colleges.* "But we are so conscious of the great value of securing your help if possible that I write at once although I am aware that you are not coming up to Oxford during this present term." The college's mathematical tutor had died; would Mrs. Nicholson agree to take his students when she did come? Grier would look for a substitute to teach them in the meantime. Dot agreed and proposed her husband as the substitute. He added this to his Balliol duties.

In the spring of 2023, John Nicholson's medical records will be opened to the public. Perhaps then we will learn what happened to him 100 years before. Nothing in the Wrinch papers hints at its nature, only that the injury was so severe that he could not finish the term. A Balliol colleague took over his teaching and tutoring. By June, John felt "nearly better."[20]

Her Girton fellowship completed, Dot joined her husband in Oxford that summer. She would, she expected, take his students in the fall. But Miss Grier had changed her mind in the meantime. The students were weak; a third tutor in two years would be too much for them. Dot could take the entering few.

Dot wrangled angrily, Miss Grier stood firm, and the tempest passed. Dot kept busy, teaching her few students, writing mathematics papers, and assisting Russell with his books. Within a few years she directed the studies of all the mathematics students at Oxford's five women's colleges.[21] Grier came to admire her. "I have seldom known a teacher who aroused so much interest in her pupils. They have again and again spoken to me with enthusiasm of their work with her, and their enthusiasm has not been confined to what they say of their work, but has extended to her wide range of interests, and her knowledge of subjects artistic and literary extends far beyond the scope of her specialized work."[22]

John Nicholson did not get better. "We were disappointed that you did not come to Toronto," D'Arcy Thompson wrote to Dot in September 1924.

---

*At Cambridge the heads of women's colleges are called headmistresses; at Oxford the title varies from college to college. The women's colleges (and some of the men's) called them principals. Balliol is one of several with masters.

"The Meeting was a great success, and the company on board the 'Caronia' was first class. We were still more sorry to be told that your good man had had an accident, and that this was the cause of your absence."[23] The British Association for the Advancement of Science had leased ships to transport its members and arranged post-conference excursions, one to Vancouver by train.

In 1970, in long-forgotten, waterlogged cardboard boxes in the cellar of Balliol, John Jones found the documents in the case. Not Nicholson's medical records—those remain sealed in the Oxfordshire Health Archives—but Balliol correspondence, including letters between A. D. Lindsay (the master) and Nicholson's doctors.

In the summer of 1924, while his colleagues sailed to Toronto, John was treated in the Cassel Hospital for Functional Nervous Disorders. "In my opinion he has been ill since 1916," wrote the medical director, T. A. Ross. (Dr. Ross did not give the basis for his opinion, but I note that John's close friend, Samuel Maclaren, was killed on the Western front that year.[24]) Dr. Ross continued, "He has had symptoms more or less continuously ever since then, but they were very much aggravated after his accident...As is so common after concussion, he has been since then more susceptible to the action of alcohol; but I do not think that there will be further trouble in this respect."

Nineteen twenty-four was Freud's heyday in London, but Dr. Ross urged his patients to look forward, not back. "Since Professor Nicholson left this Hospital on August 5th he has tackled some very important family business which had been hanging over his head for many months and which he had felt totally unable to approach. I consider that his having done this is a very important indication of his recovery." (Again, the doctor gave no specifics.) He pronounced his patient "perfectly fit to return to work with safety to himself and to others."

There the Balliol record falls silent for six years.

Classics was to Oxford what mathematics was to Cambridge. Through the centuries, the two universities were the opposite poles of British highest education. When Dot moved to Oxford in 1923, she entered a different world, a world not ruled by the Mathematical Tripos. But she was not isolated professionally. Hardy had recently moved to Oxford too; in the wake of Russell's sacking, he had accepted the Savilian chair in geometry. E. A. Milne, once her fellow Cambridge student, held another of Oxford's few mathematics chairs. The eminent, elderly A. E. H. Love of the seismic waves had been there all along.

"Mrs. Nicholson by the force of her personality and her unsparing devotion is a great asset to mathematics in Oxford," Milne wrote in 1930. "Neither her research nor teaching cover the whole ground of her mathematical activity. She takes a prominent part in affairs of organisation, of curriculum, of examinations, and of the development of policy. I think it is true to state that her judgment in these matters is much respected." And no one, Milne continued, had published papers in so many areas of mathematics: logic, set theory, asymptotic expansions, relativity theory, hydrodynamics, electrostatics, aerodynamics, and other areas of mathematics and mathematical physics, and the philosophy of scientific method.

Dot's resume reads like a Pirandello play, fifty answers in search of a question.

All that, and morphology too.

"I am working out some new wing profiles suitable I hope for aeroplanes," Dot told D'Arcy Thompson, "and I want to find out if there is any bird whose wing profile is of a similar type. Aerodynamics is peppered with albatross wing sections and the herring gull etc. I want to find some type of wing profile sufficiently similar for the loan of the name of the bird to the section to be suitable."[25]

Obviously, she had never examined a bird. "As to your Birds' wings," D'Arcy replied dryly, "I should like to take you to Leadenhall market: where, any day of the week, you may find birds of many sorts, and study their wings to your heart's content. The curves you draw seem to me excellent curves, but not a bit like the profiles of any real bird's wing:—i.e., if I understand what you mean by a 'profile,' which I should rather call a section of the wing."[26]

But no other mathematician was so interested in biological form. "There are heaps of questions I should like to ask you," he continued, "about stream-lines, and various bird and fish-forms: in fact, you are going to have a few such questions before long. There must, I suppose, be a simple answer to the question of why so many birds' tails end in a long fork—swallows, albatrosses, etc., etc."

Dot guided D'Arcy to his answers. He guided her to her question.

10 HOMES ARE HELL

Pamela Wrinch Nicholson's birth in London on March 20, 1927, went unrecorded at Oxford. No college's council minutes logged requests for time off, not before and not after. Miss Grier noted nothing at Lady Margaret Hall. Pamela's "godless godmother," Margery Fry, left nothing in Somerville's jumbled archives. Not a clue at St. Hilda's, or St. Anne's, or St. Hugh's. Only this footnote in *The Retreat from Parenthood*, a cryptic, pseudonymous, but revealing book Dot wrote two years later, alludes to this pivotal moment in her life:

> It is interesting, historically and anthropologically, to observe:
> a) That in at least one women's college there is a statute prohibiting the employment of married women so long as their husbands survive. b) That two University teachers, recently the recipients of a graciously given *permis de marriage*, found that a *permis de reproduire* was not forthcoming when they applied for it a few months before the birth of their child, staged for the middle of the long vacation.

Dot faced down motherhood like the Tripos. In her book, the pseudonymous author "Jean Ayling" describes a visit to "Angela," a stand-in for Dot. The young chemist

> for eighteen months past has lived a life which, apart from her professional activities, has been bounded on every side by textbooks of mothercraft, textbooks of dietetics, textbooks on every subject in the world that could possibly have any relation to the welfare of her young child, *in utero* or in the outside world.
> It was quite overpowering sometimes to come unexpectedly into her study and find her immersed in "separates"—which

she had extracted by the simple method of writing to their authors—
"separates" of every kind, and in many different languages, from peri-
odicals and journals published all over the world....

I saw her applying exactly the same standard of workmanship to
her researches in maternity as she habitually used in her profession.

The Retreat from Parenthood was published in 1930 by Kegan Paul in its
series Books on Sex and Morals, but sex is mentioned nowhere in it, and
the only moral question is how to save the British race. Jean Ayling did not
flinch from the hot topic of interwar England: why is it that "during the
last twenty or thirty years, doctors and lawyers, scientists and civil servants,
engineers and educationalists, artists and architects, and all the rest of the
men and women who earn their living in professional fields, have so severely
curtailed their fertility?" In searching for answers, Dot drew on contemporary debates over women's rights and roles, experiments in schools, children's
camps and orphanages in the new Soviet Union, the eugenics movement, and
the uncertain trumpet of science.

The narrator struck the pose of a supercilious, disinterested observer. "In
the case of many people the greatest—perhaps the only—service they can do
mankind is to remain barren"; yet she "found it painful to follow the decline
of breeding" in a group of men and women that "stand as good a chance
as any of producing sound descendants." The British Empire teetered on the
cusp of collapse; its future, if it had one, must be shouldered by those trained
to lead it. But the ignorant were breeding like rabbits on the downs, while
those who *should* be breeding—the professionals, the educated—were choos-
ing to be childless. What is to be done?

The Nicholsons lived in the leafy village of Woodcote, midway between
Reading and Oxford. I took a bus from Reading, got off at the Woodcote
post office, and asked for directions. Dot's letters from that period give no
street address, only "Beechwood," the house's name. The immigrant postmas-
ter had never heard of it, so I set off on foot. An elderly native collecting
trash with a long-handled pole in a grassy park didn't know either. The par-
ish clerk would, he said, but he's only in on Wednesdays. Maybe it's on Beech
Lane. He pointed to a shady road winding up and away from the village.

Some houses on Beech Lane bore name signs; I found a "Beechcroft" and
several other Beech-compounds, but no Beechwood. Two workmen adding
a new bathroom to an old whitewashed duplex told me to knock on the
door and ask. When I hesitated, they cheerfully did it for me. The owner,
a friendly, middle-aged man with graying hair, didn't mind the interruption.

Beechwood would have been an old duplex like this one, he said. There are several on this street, all built around 1890. The other houses are postwar.

I'm writing about a woman who lived at Beechwood in the 1920s, I told him. She wrote a book on parenthood; one chapter title is "Homes Are Hell." *That's right!* he exclaimed. Homes *were* hell in those days! His got its first and, until now, only indoor bathroom in May 1930; he found the date on a crumpled newspaper wadded between the studs of the old bathroom wall. The lead story was the British prime minister's curt response to a French proposal for a United States of Europe: "Not for England."

Jean Ayling wrote grimly of outdoor bathrooms, burst mains, chimney pots blowing off the roof, warped gates, keys that didn't work, doorknobs that fell off, and cracked walls. The little laundress in the next village gave up, and sour milk was delivered by boys coughing as they ladled it. "Is it an exaggeration to say that the home is hell?" And, she added, this will seem like nothing when the children come.

A proper home should have hot and cold running water and proper drainage, soundproofed and washable walls, tiled floors in the kitchen and bathrooms, windows made to open "and so hung that both sides can be cleaned without danger to life." Gas for cooking and heating, and electricity for vacuum cleaners, sewing machines, and other modern labor-savers. Balconies or verandas for sun and air bathing, a garden of at least a quarter acre per household of five people, and access to a swimming pool and a gym. But English homes had none of these. That's why women with children were exiled from their professions. That's why women in professional fields paid the price of sterility.

In a series of vignettes, Jean Ayling yanked open the curtains of England's happy homes. One Joseph Brown, B.A., comes home after a long trying day in the office to find his wife and child slumped on the couch. Instead of bringing joy, "the various tiresome, tedious and time-absorbing things" that motherhood demands have left her pale, listless, and utterly miserable. Mr. Brown cares. "'There's not much a fellow can do, you know,' he says, 'but look here, darling, what I have for you!' and from his neat attaché case he brings out some Fuller's chocolates and a copy of *Good Housekeeping*."

Muriel Wrinch is not cited, though Dot's sister's articles in *Good Housekeeping* were internationally acclaimed. The journal, founded in 1885 in Holyoke, Massachusetts, a few miles from Smith College, preached healthier living conditions, better nutrition, and enlightened mothering. It reviewed new drudgery-reducing appliances and gave its Seal of Approval to those that worked best. A British edition was started in 1922.

Muriel already had two books to her credit, *Mothers and Babies*, written with her husband, Helmut Schulz, on their honeymoon in 1924, and

Your Children, published two years later. She argued for modern mothering: educated, science-literate, involved. Motherhood is "a profession no less than medicine or engineering," she wrote. It must incorporate the latest findings in prenatal training, postnatal training, training in bathing, breast-feeding, hygiene, child development, management, music, and storytelling. *Mothers and Babies* is dedicated "to one mother [hers] who always had faith in the work of her children, and to the memory of another [his] who made of her motherhood a profession."

After Surbiton High School, Muriel had enrolled in medical school at St. Thomas Hospital in London. There she met Helmut, a son of German Lutheran missionaries in South Africa. Muriel switched from medicine to early child development, earned a higher Froebel certificate (then the field's top credential), wrote books and articles, and lectured on the BBC.

Dot and Muriel agreed that motherhood is a scientific profession and that women should be thoroughly trained for it. They disagreed on *which* women.

"The keenest joys are those which result from fulfilling the purpose of one's life—from fitting into one's place in Nature," wrote Muriel. "Woman is created to be a mother...Nearly all women can, with suitable training, become mothers and it is to the advantage of the community that all women should so train themselves."[1]

"Motherhood as a full-time employment has failed," wrote Jean Ayling. "This at least is the opinion of the average young woman who, in possession of a healthy body and an intelligent mind, looks forward to a life of professional activity."

Drawing on her formidable powers of synthesis, Dot distilled the myriad separates of every kind in many different languages into a proposal for a national system of child-care support. Brushing aside the "outmoded, mawkish sentiment" that "from the womb to the school door, the needs of the child must be supplied within the family," she called for a national Child Rearing Service.

The CRS would have four specialized branches. Branch A would redesign homes and the appliances therein. Branch B would train skilled domestic and child-care workers. Branch C would organize cooperatives for purchasing and distributing wholesome food. Branch D would "cover all the medical, nursing, and auxiliary services which are concerned with the physiological and psychological health of the pregnant and lactating woman, the embryo, the foetus, the infant, and the pre-school child." With the CRS on call, the new mother could pick up her golf clubs and tennis racket, drop off her baby at the Home where she was born, and go on vacation in Switzerland. "You will, I know, ring up each day," the fictional chemist, Dot's stand-in, tells the nurse, "and you *do* know my telegraphic address, don't you?"

An all-encompassing, revolutionary organization like this could not spring battle-ready from her head, Dot knew, but there were precedents. "Sixty years ago the education of girls was in a condition similar to that of child rearing today. A few people here and there had views as to the desirability of giving education to girls. But there was no detailed scheme, and no finances were available." Like-minded citizens had banded together and established the Girls Public Day School Company.

"Dear Dot, I have now finished your book on *The Retreat from Parenthood* and I think it a quite extraordinarily good book," wrote Russell.[2] "Your proposals in the second part are very constructive, and I wish I could suppose that they would be carried out. As a reformer, however, I have become somewhat old and disillusioned, so that many things which I know should be possible, appear to me very difficult." He continued in this discouraging vein. "Take, for example, your C. R. S. Have you considered what would happen if such an institution really existed? It would, of course, have Presidents and Vice-Presidents who would be titled busybodies, anxious to see that its activities were all points in accordance with the teachings of the Church of England." And so on. "I think the book very able," Russell concluded, and "I shall do all I can to make it known. Are you very anxious to preserve the anonymity strictly, or may the authorship of the book be mentioned privately?"

Dot dedicated her book, mysteriously, to "my friends M, S, and R." R is certainly Russell. If S is Sara Margery Fry, then M might be Muriel. But friends called Fry Margery; in which case S is a mystery.[3] As for Jean Ayling's identity, Dot had good reasons not to bruit it about.

First, the jack-of-all-trades label. Dot's friend, the geneticist Lancelot Hogben, sat on his *Mathematics for the Million* (still in print today) until his election to the Royal Society vaccinated him against that charge. But even without *The Retreat from Parenthood*, Dot's unfocused resume was a concern. Hardy and Milne urged her to pick one field in mathematics and plow it thoroughly. To reveal herself as Jean Ayling would be the last straw; it would wreck her mathematical career.

But that argument assumes Dot was savvy, which she rarely was. There is a simpler explanation, and I find it more probable: she did not want, or her parents did not want her, to compete publicly with her sister.

The sisters disagreed on who should practice motherhood, but not on how to practice it. "*Regularity* is of the greatest importance with the human baby. It is the beginning of his discipline," Muriel explained, and Dot agreed. "The baby is to be taught that sucking, although an invaluable means of obtaining food, is not to be indulged in indiscriminately." He must be fed promptly at 6 a.m., 10 a.m., 2 p.m., 6 p.m. and 10 p.m., and never in between: "even if the baby is asleep at his feeding-times he must be awakened. For if the mother hurries to the baby whenever he cries, she is making him into a self-ish little tyrant who has only to raise his voice in temper to be instantly obeyed....Do not rock him, do not take him into your arms and carry him up and down...after ascertaining that the child is not uncomfortable, she must leave him severely alone." Behaviorism was in the air, the air exhaled by the American psychologist J. B. Watson. The fumes lingered for decades until Dr. Spock blew them away.

When her daughter Talitha was born in 1926, Muriel practiced what she preached. The Schulzes lived, at first, with Muriel's parents in Surbiton. Each day, Muriel parked Talitha in her pram in the warm sunny garden, with mittens on both hands to keep her from sucking her thumbs. Each day, Grandma marched out and pulled the mittens off.

But the little family soon moved to South Africa and settled in Robertson, near Capetown. Carla, Peter, Rosalie, and Christopher were born in successive years. "Every Wednesday the mail arrived by steamer," says Talitha, "and each week there would be something from grandma for one member of the family, you never knew for whom it would be. She didn't send the presents in order, 1,2,3,4; you never knew. I loved that." But she never saw her Wrinch grandparents again.

Russell did not return to the contemplative life. John, his first child with Dora, was quickly followed by Kate, and houses in London and Cornwall added to the bills. So Russell wrote books that would sell—books like *What I Believe, Why I Am Not a Christian*, and *Marriage and Morals*. His colleagues were predictably unimpressed. Dot thought he was wasting his intelligence, but this did not affect their friendship. Letters and pocket diaries, his and hers, show they lunched together off and on through the years when they both remained in Britain, into the mid-1930s. In 1929, the Russells amended their wills to appoint Dot their children's guardian.

To raise John and Kate "free and fearless," the Russells opened a school of their own. Telegraph House, with its many rooms and vast woodlands, was the perfect setting for it; reluctantly, Russell's brother Frank agreed to rent it to them. He "had speculated unwisely," Russell explains, "and lost every penny that he possessed. I offered him a much higher rent than he could have obtained from anyone else, and he was compelled by poverty to accept my offer. But he hated it, and ever after bore me a grudge for inhabiting his paradise."[4] The Beacon Hill School was progressive and coeducational, with no prudery, no religion, and no required studies (but the children had to brush their teeth and go to bed on time).

Unexpected expenses, problem children, insufficiently dedicated staff: the school was soon more hell than home. But neither Russell nor Dora let this interfere with their writing and lecturing (Dora for the birth-control movement). They traveled, separately, for months at a time. Nor did it cramp their style. Kate, as a child, thought all daddies napped with the nannies. Dora fired a cook for reporting an instance.

Dot's list of published papers grew ever longer (for an assessment of her work by the French mathematician Jacques Hadamard, see the appendix). In the spring of 1929, she applied to Oxford's Faculty of Physical Sciences for another doctor of science degree and became the first woman in Oxford's history to receive one. The judges wrote:

> As regards the nature and quality of the work we agree in stating that the fifteen papers submitted constitute a contribution to the mathematical sciences which is both original and important...The subjects treated cover a wide range in both mathematics and mathematical physics. Some of the papers contain distinctly novel and interesting ideas, others are occupied with the systematic development of these ideas, others again are concerned chiefly with pointing out applications to practical questions. So far as we can judge, all the work has merit, and some of it is of exceptional quality, breaking new ground and exploring very completely the possibilities opened up by the new ideas that are originated.[5]

One of those exceptional papers had appeared the year before in the *American Journal of Mathematics*,[6] a phoenix arisen from the ashes. With Hardy's encouragement, Dot had sent him a manuscript on asymptotic expansions for the *Cambridge Tracts*, but he had rejected it. "She has omitted much that ought to appear, and the portion of the subject she has dealt with is far from being the most important," the referee wrote. "The method is certainly an ingenious little device of some practical importance for finding asymptotic expansions of certain classes of functions. But these classes are very limited." The *Cambridge Tracts*—the series continues today—are overviews of specialized mathematical fields written for nonspecialists; Dot's manuscript was too idiosyncratic for such an audience. After criticizing several technical details, the referee added, "Finally, even if the contents of the manuscript seemed commendable, I should have to advise the Press to request the author to modify the symbolism."

Other rejection letters said much the same.[7] The editor of *The Proceedings of the London Mathematical Society*—her former tutor and perhaps spurned lover, G. N. Watson—wrote of one of them, "The main points which the referee stresses, not only in the detailed report, but also in a covering letter to the Council are as follows: He considers the main idea interesting, but that it has limitations which you do not indicate, and in fact you ignore them when they are not to be ignored; and that the writing out of the paper is sketchy, and the claims which you make in it are exaggerated." Speaking for himself, Watson added, "I was perpetually being pulled up with a sense of irritation at the general style of the paper." It was published later in another journal.[8]

In an undated report on an unspecified paper, another referee wrote, "In my opinion the author would be well advised to abate the somewhat exaggerated claims which she makes as to 1) generality, (2) simplicity, in [sections] 4,5,6 and on pp 19,20 and in other places." Next to this, Dot penciled, "I don't agree."

Criticism, rejection, and revision are the electrostatics of academic life. Mathematics editors send many, if not most, papers back to their authors for revision. The authors argue, or comply, or turn to other journals. Dot did all three. (Some authors sulk; she never did.) These reports, painful to Dot, are gold mines for me. They preview weaknesses that would plague her later, when she ventured into biology and chemistry. Not errors in logic, but errors in judgment.

Motherhood did not slow Dot down or confine her geographically.

In 1928, Oxford sent her and Hardy to the International Congress of Mathematicians in Bologna. It was hot there in August; she relaxed at the hotel swimming pool. Bertha Swirles (the future Lady Jeffreys), also a mathematician by then, lolled about with her. "Nicholson sent her a daily telegram to the effect that Pamela was optimum or super-optimum," Bertha recalled. "They were still happily together at that point."[9]

In 1929, the British Association for the Advancement of Science met in South Africa. The 535 visiting overseas members sailed to Capetown on three ships. Pamela stayed behind.

The meeting opened in Capetown on July 22. Most of the sessions were for the attending scientists, but D'Arcy Thompson gave a public lecture on "Anatomy from an Engineer's Point of View." Dot visited her sister, who lived nearby, and met her young nieces and nephew. Perhaps it was on this

Dot Wrinch, standing, with her sister Muriel and Muriel's children, Talitha, Peter, and Carla, at their home in Robertson, South Africa, July 1929. Photograph courtesy of Julian Williams.

visit that she brought them the kaleidoscope they remembered all their lives. At the end of the month, after an excursion to Kimberly to see the diamond mines and "other features of interest," the entire delegation traveled to Johannesburg for another round of scientific lectures, including one by Dot. More excursions followed and then they all sailed home.

Her husband was imploding, but Dot kept the curtains of her own happy home drawn tight. She told no one, not D'Arcy, not Og, not Russell, about John's drinking and outbursts. Nor did she tell anyone she was plotting to leave him. A footnote in *The Retreat from Parenthood* mentions an enlightened venture: Smith College, in Massachusetts, United States, had established an Institute for the Coordination of Women's Interests, charged with finding "practical ways to make household chores easier for women and to educate women to pursue interests which could be integrated with the duties of marriage and family." Its notion of integrable interests was limited to domestic and landscape architecture and freelance writing, but the institute was in the forefront of household management. In just three years, it had organized a cooperative nursery school, arranged for cooked meals to be delivered to faculty women's homes, and trained young women to serve as household and personal assistants. At Smith College, perhaps, Dot could teach mathematics and raise Pamela too. But, the dean replied to her inquiry, the mathematics department had no openings nor were any foreseen.[10] Behind the scenes, the institute was plagued with financial and organizational problems; it closed within the year. When Dot did come to Smith, in 1941, it was a fond but receding memory. When I joined the faculty a quarter century after that, the small band of emeritae female faculty still spoke of it wistfully.

Dot's friends may not have been aware of John's deteriorating condition, but his students certainly were. Reginald Victor Jones, the future "secret hero of British intelligence in World War II, whose genius for scientific tricks baffled German bombers and saved thousands of lives,"[11] never forgot Nicholson's course on hydrodynamics in Hilary Term, 1930. Five students came the first day. The next day R. V. was one of two, and the course was canceled by mutual agreement. The professor "made an indelible impression," R. V. told John Jones. He appeared in a tattered gown and lit up one cigarette after another, though smoking in academic dress was forbidden. Then, "saying that his gums hurt, [he] proceeded to remove his false teeth which he placed beside him on the rostrum. The resultant slurring did nothing to improve the incoherence of his words, and so we learnt little."[12]

"The story among us undergraduates," R. V. continued, "was that he had been a truly distinguished mathematician who had been disappointed by being 'scooped' by Planck in formulating the quantum theory, and had ever afterwards drowned his disappointment by reverting to the Balliol cellars." The students had the story wrong: Nicholson wasn't scooped by Planck, he was trumped by Bohr, years before Balliol lured him to Oxford from King's.

In the spring of 1930, John was treated for alcoholism at the Norwood Sanitarium. "His addiction is of a very mild type, his wish to be cured is quite decided," his doctor assured A. D. Lindsay, the master of Balliol. "Alcohol does not suit him at all, and he should never touch it again. This he now well understands. It has been his misfortune to tamper [?] with alcohol, without understanding that he belonged to a type unable to take it without risking spectacular trouble. Many people can take a much greater quantity without anyone knowing anything about it. With Dr. Nicholson it would soon be obvious."

The doctor's gingerly optimism notwithstanding, Dot had had enough. She and John signed a legal separation in the summer.

Soon afterward, though he had not touched a drop, John's behavior became unstable and alarming. "I spent an unpleasant 3 hours or more with him first in the afternoon and then in the evening," another doctor wrote, "and Dr. Neill [the superintendent of Warneford Hospital] had a worse time than I had, as he coaxed the patient into a taxi and got him off without any disturbance."[13] The doctor enclosed a bill for £3.30 for three hours spent with the patient, for signing committal forms, and for an "interview with the Police."

At Warneford, John Nicholson was certified a lunatic. He remained there until he died in 1955.

Concussions "can lead to long-term problems," the National Football League admitted in 2009. This should not have been news. A 1928 study "methodically details the well-publicized problems—loss of coordination, cognitive deficits, uncontrollable rages."[14] Did the concussion John suffered in 1923 lead to his dementia? The Balliol papers do not say, but neither do they ascribe his dementia to alcohol. The alcoholism was a secondary result due to a more fundamental instability of mind, Sir Farquhar Buzzard, Regius professor of medicine at the University of Oxford and physician-extraordinary to King George V, told Lindsay, the master, after examining John.[15]

Lindsay evidently agreed. He went to great lengths to ensure John's long-term care. Though Balliol decided not to renew John's fellowship when his term ended, it agreed to pay for his "care, maintenance, and well-being" at Warneford as long as he remained there.[16] "Lindsay was a strict moralist," says John Jones. "If he thought Nicholson had brought this on himself, he'd have let him rot."

Letters in the waterlogged Balliol cartons suggest Dot walled off her emotions in response to the tragedy. "With regard to the books belonging to the patient…the range of scientific interests of the patient was very wide. With regard to the furniture it is little value and should be sold with the exception of the small cabinet and arm chair and gold bag and walking stick which it would be a pleasure to the patient to have sent to him I should imagine." Dot gave her husband's butterfly collection to the university museum and selected books from their large collection to keep for herself. *Insect Architecture, Secrets of the Totem, British Reptiles, Architecture and Painting, The Rape of the Lock*, Milton's poetical works, Shakespeare's plays, *The Theory of Sound, Volume I*, Irving's *Sketchbook, My Life and Adventures, Spirals in Nature and Art, Microscopic Funghi, Don Quixote, The Life of Johnson, Relativity, Social Life in Egypt, Electricity and Magnetism, Le Kama Soutra*, and Dante's *Inferno* were a few.

"I have written at once to Dot," Dora wrote to Russell in July, "asking her to come see me here [in London] and suggesting Cornwall in early August, because she must be feeling wretched. I did not say anything though about her marriage—just told her the baby had arrived, etc."[17]

But the Russell household was growing ever hellish. This baby was Dora's third, but not Russell's, nor would her next child be his. Though theirs was an open marriage, the two external babies and Russell's "transfer of affections" to a young woman working in the household increased the tension. His brother Frank died suddenly in 1931, leaving Telegraph House and everything else to one Miss Otter. The lease had six years to run; the school could stay. Russell chose not to. Dora managed to keep the school going, but John and Kate were unhappy there, and Russell took them out.

Dot's sister Muriel found the grass no greener at the other end of the earth. As a British citizen in the Boer community, she was isolated by language, by culture and, later, by her outspoken opposition to apartheid. Helmut's father had never welcomed his "fremdling" daughter-in-law. Though Helmut had a busy medical practice with white, black, and colored patients, the Schultzes lived on the margins. One patient paid him with a monkey; it ran up bills stealing fruit from a nearby shop. Another patient paid with a parrot that cursed when water was sprinkled on its tail; somehow this always happened when the parson came to call. "Neither of my parents had any money sense," says their daughter Rosalie. The marriage collapsed in 1937, when Helmut ran off with the children's piano teacher. Muriel stayed in South Africa; raised the five children; lectured widely on child care; and ran a money-losing boarding school where the children ran naked, the girls did carpentry, and the boys learned to sew. She "was an intellectual, not a motherhood queen," Rosalie continues. "She lectured in Rhodesia and South Africa, teaching the modern ways. Her determination to cultivate her interests outside the home contributed to the breakup." Dot sent money to help her get by.

11 METAMORPHOSES

"With regard to J, all is tragedy," Dot wrote to Russell. "I can hardly bear to think of it, but there was no other way. But it is fearful when one thinks of his work. It is an awful grief to me to think of a good mathematician going so utterly to pieces. But *was zu tun?*"[1]

Was zu tun, what to do: get out of Oxford; get a fellowship for study abroad.

Oxford dons, both men and women (said the Rhodes Foundation circular), were eligible for traveling fellowships to learn about the countries and backgrounds of Oxford's Rhodes scholars. Dot applied to visit universities in America ("at least one State University, one University privately controlled and one Women's College") to learn how higher education was organized there. G. H. Hardy, Lynda Grier, Margery Fry, and other Oxford luminaries wrote enthusiastic letters of support. "Her passion for work is equal to her ability," said Fry.[2] "She has in many ways the most remarkable intellect I have known in a woman during a lifetime spent largely in the company of academic women in this country."

The Rhodes Foundation turned Dot down. Hardy protested to its Oxford representative, the warden of New College. But of course! the warden explained. We cannot give her a fellowship. Returning Fellows are expected to mentor the Rhodes scholars at their colleges. Since the scholars are men and only men teach them, the foundation cannot, syllogistically, give a traveling fellowship to a woman. Hardy put up a fight: "I got at the Warden on Friday and had a long argument with him, but a very unsatisfactory one," he told Dot. "My disappointment was that I just failed to get any sort of sympathy out of him. He did seem to be definitely anti-feminist…our beloved Warden did put my back up very much, and the less he wants you to have one the more I do. But it's no earthly good my resuming the topic directly with him."[3]

Dot reapplied the next year. Again Hardy supported her. The warden didn't budge.

Germany, then. The Rockefeller Foundation's International Education Board offered fellowships in science (mathematics included) "to assist young scientific men who are working under the directions of scientists. Grants will be made to men of unusual promise in their respective fields so they may pursue abroad, under guidance, studies which they can not pursue at home with equal advantage."[4] Women were not excluded by the "he," but very few received them.

Technically, a candidate did not apply for a Rockefeller fellowship; instead, eminent specialists applied on his behalf. Professors Hardy and Love applied for Dot to do research in mathematics at the University of Göttingen, the Cambridge of Germany where mathematics was concerned. Miss Grier applied for her too, but in the Social Sciences Division, to study how professions were organized in Germany. Fatefully, Dot neglected to tell Hardy and Love about the sociology proposal, nor did she tell Grier about theirs.

Hardy judged the Rockefeller a very long shot. Dot's varied accomplishments did not add up to any one big thing; the breadth of her research would count against her. Dot argued back: tutoring and mothering and dealing with John had left her no time for big projects. With a fellowship, she would show what she could do. Yes, said Hardy, but "the people I have recommended before have all been 100 percent specialists in analysis. In every case I have been able to say—there are the works, perfectly clear cut, just what I say they are, and I can challenge any expert to say the contrary...Take Zygmund, for example: he knows the theory of trigonometrical series better than Littlewood and I do."

Nevertheless, Hardy helped her every step of the way.[5] Get a letter of support and sharpen your proposal, he advised. "It is...no good saying that you want to learn a subject in the vague hope that you might find something in it to do." In another letter he warned the perennial wrangler: "It is of course not unnatural that they should jib at paying a stipend for study in Göttingen over a [vacation] period when you obviously could not do so effectively. It would be fatal to put their backs up at the beginning by trying to dispute this."

Then word of the other application reached him.

"I have had a letter from Dr. Tisdale [head of the Rockefeller Foundation's office in Paris] to which I have found it extremely difficult to make any satisfactory reply," Hardy spluttered:

You never warned either Love or me that you were proposing to make a second application in an entirely different field. We put the case (as I told you I was bound to) on the ground that, while it was impossible to put the work you had actually done on a level with that of (say) Pólya or Ostrowski, there were considerable possibilities if only you had a chance of turning your self into a "100 p/c" mathematician in a definite field; and that that was what you were determined to do. It was the only way of putting forward the application with a one in twenty chance of success. Now Dr. Tisdale retorts that apparently you are equally willing to desert mathematics altogether— and what on earth is the reply? I have replied as best I could, but he has us waving our bats vaguely—it looks as if neither you nor Love nor I had any conception of what the standard of the fellowship is; and, however one looks at it, the chances of success are very gravely prejudiced.[6]

Dot apologized to Love, with a copy to Hardy, but stood her ground: "I am determined to become a proper mathematician with a definite field.... Tisdale cannot properly deduce from the existence of the other application that I am 'equally willing to desert mathematics altogether.'...I know it is odd to have a serious interest in two fields, but I just have it. I have always had it." Love replied graciously; Tisdale had agreed to bring her application to the board. To everyone's disappointment but no one's surprise, the board said no.

I studied the minutes of that board meeting, now in the Rockefeller Foundation Archives at Sleepy Hollow, New York. I was puzzled to find that Dot's application wasn't mentioned at all. A note in Tisdale's diary suggests he never intended to bring it up. "That's how he would have handled it," the archivist told me. Tisdale's deception turned out to be a kindness. Five years later, when the board considered her for a research grant in biology, there was no rejection on the books.

The social science application failed too, but not before another flap. Alexander Carr-Saunders, an eminent sociologist soon to lead the London School of Economics, had written a book on the professions—Jean Ayling had cited it in *The Retreat from Parenthood*—and he was working on another. Dot proposed they collaborate. It's too late, he told her. His new book was nearly complete.[7] She didn't or wouldn't hear him; she told Grier, with great excitement, that he had agreed they would work together. A perplexed Carr-Saunders got wind of this and complained to Grier. The soul of discretion, she invited him to tea.

Drawing on her organizational work for the National Union of Scientific Workers, Dot had proposed to study "the conditions of entry into and expulsion from the professions" in Germany; "training, salaries and employment conditions; opportunities for in-service training," and "the attitude of established practitioners toward free access to the profession of those with the necessary intelligence and training." She would deal with these matters for the rest of her life, but as a participant, not an observer.

Girton College came through for her again, this time with a Hertha Ayrton Fellowship. Dot could study mathematics in Göttingen after all. She "imported a German" to Beechwood to immerse herself in the language and wrote cheerful notes to Russell and Og. Then the other shoe dropped. By the terms of the legal separation, Balliol deposited a portion of John's salary in Dot's account each term, but as an inmate of Warneford, he drew no salary. Girton's fellowship would not stretch to cover travel and living abroad and the Beechwood mortgage too. Russell tried, but failed, to interest an American publisher in *The Retreat from Parenthood* to help her earn money.[8]

So Dot taught in Oxford for one year more, in which she sold Beechwood and paid off the mortgage. The next year, Lady Margaret Hall gave her a fellowship to supplement Girton's. Dot set off, Pamela in tow, in August 1931. Not for Göttingen, but for Vienna: the city of chocolate, opera, and 1,000 coffeehouses where mathematical philosophers, philosophical mathematicians, and like-minded biologists sat at nearby tables, talking into the night. Sometimes they sat at the same table.

The reactions of Hardy, Milne, and Love were not recorded, but they must have been stunned. They had helped Dot acquire letters of welcome from Edmund Landau and Richard Courant, leading mathematicians at Göttingen, and she had drawn up an impressive work plan. But other forces were tugging her to Vienna.

Vienna was home to the Vivarium, a famous biological laboratory. Hans Przibram, a biologist, was its head and his physicist brother Karl one of many scientists and philosophers who argued at its tea table. Hans studied the growth and form of insects; Karl was an expert on Brownian motion. Where D'Arcy Thompson saw analogies, the Przibram brothers did experiments. Where he found elegance and simplicity, they found chaos and complexity. Where they glimpsed biological laws, he suspected leaps of imagination. But D'Arcy and the Przibrams were allies in the nascent international campaign to infuse biology with physics and chemistry.

A friend at the Aristotelian Society, Joseph Woodger, a Middlesex Hospital biologist, had gone to Vienna a few years before to do transplant experiments on earthworms. Instead, he talked philosophy day and night and returned an evangelical bio-logician. His friends began to call him Socrates and so did he. In his debut lecture to the Aristotelians in 1929, he chided scientific methodologists who thought biology was "too immature to off er much scope for their talents." On the contrary, he said, biology was ripe for restructuring. Perhaps in response, Dot lectured that year on "Scientific Method in Some Embryonic Sciences" (pun evidently intended).

D'Arcy may have urged her to go. We know he urged her toward biology per se. Start with cell division, he told her. "Investigate the field of force within the dividing egg (or other cell); the conditions of stability of its surface-equilibrium and the point at which the unstable equilibrium breaks down, and is replaced by the new stability of the divided cell."[9] All her work up to now, electrostatics, hydrodynamics, aerodynamics, scientific method, and her talents and drive and ambition, would find their raison d'être here. Scientific method and mathematics, the two streams of her ever-whirring mind, would bond in this sticky biological problem.

"Nothing is more wanted in biology than that mathematicians of first-class standing should interest themselves in biological problems," D'Arcy told the University of London, hoping to tap its Dixon Fund for her. "I do not know of anyone better qualified—or even so well qualified—as Dr. Wrinch to undertake the task. I strongly recommend that she be helped and encouraged to apply herself to it."[10]

The Dixon Fund chose not to help her, but Dot plunged in anyway. "I have just been to see Gray at Cambridge," she told D'Arcy ("my most kind and charming Patron").[11] James Gray, F.R.S., but not yet knighted, was the author of the pathbreaking *Text-Book of Experimental Cytology*. Gray used cinematography and the stroboflash to track the dividing cell. He lamented the lack of a theoretical framework, a logical structure, for interpreting his findings. D'Arcy encouraged Gray and the growing group of English experimental biologists. Regrowing, better said: Their little band, all but one, had been killed in the Great War. It was time that form be studied "from the side of chemistry and physics, with the aid of the known properties of matter and energy."[12]

"Any knowledge of the distribution and magnitude of the forces likely to exist within such systems would form a most valuable guide for future experimental work and for a proper understanding of the underlying mechanism," Gray said. "The problem is of very wide interest but requires for its solution a mathematical technique."[13]

"We are going straight ahead with plans for the great research!" Dot crowed. (This time, it was true.) "WHAT, I ask you, would one have done without your book to inspire one all this time?" In *On Growth and Form*, Dot marked up D'Arcy's chapter on the structure of the cell, noting with approval his remarks on electrical theories of mitosis.

Vienna, in the fall of 1931, was not a relaxing place to be. On October 3, the Austrian National Assembly passed an austere economic program to balance the budget and qualify for a loan from the League of Nations. "We feel very much in the middle of things here," Dot wrote to Grier, "with the little Fascist Revolution the other day and so many exciting happenings, but I do wish I could see what IS going to happen. Don't you think that this feverish economising is the straight road to a final break up of the present system and where will we then find a new system which is neither Fascism nor Bolshevismus???? I can't help wondering and wondering."[14]

The Vienna Circle didn't meet in a coffeehouse; it met in a dingy basement room on Boltzmanngasse. Did Dot visit the basement? The Circle's records don't say. But she became friends with three promising young regulars, Kurt Gödel, Olga Taussky, and Karl Menger. Gödel had announced his Incompleteness Theorem the year before, shattering Russell's program of reducing all mathematics to formal logic. Olga Taussky, born in 1906 in what is now the Czech Republic, had not made her mark in mathematics yet, but everyone knew she one day would. Karl Menger catapulted to fame in 1927 when, as a 25-year-old student, he had solved a famous mathematical problem: to find an all-encompassing definition of the mathematical notion of a "curve." The real question, Menger told his professor after a few hours' thought, is not "what is a curve?" but "what is dimension?" and handed him the definition used today. In 1931, Menger was a leading light in topology and led a lively Mathematical Colloquium at the University of Vienna.

Topology, a growing field little taught in England, was just what Dot needed to learn. D'Arcy hoped it could describe the changing shape of the mitotic cell, the branching veins in insect wings and leaves of plants. This flexible geometry could be the language of growth and form, he thought.

Which ideas, images, impressions penetrated Dot's Viennese cocoon, what guided her ineluctable metamorphosis? In the eight months between August 1931 and April 1932, Dot rearranged herself into a—a what? "Mathematical biologist" had not yet been coined.[15]

Dot consulted both Przibrams, yet she never mentioned their laboratory in her letters to D'Arcy. Did she attend Menger's Colloquium? Like the Circle's, its records are silent. Later she said she studied geometry with Menger for a year, but Franz Alt, Menger's student at that time, remembered a visit of weeks, not months. "It may be an overstatement to say that Ms. Wrinch 'studied' with Menger," he told me, "more that he listened to her ideas. I think he took her quite seriously. I also remember something about the geometry or topology of the veins in the leaves of trees and other plants."

But Alt was not always with his teacher when this particular visitor was. Karl is a clumsy lover, Dot told Olga.[16]

"I think I have got something about venation in leaves at last!" Dot wrote to D'Arcy from Vienna. In April, she parked Pamela "bei meine mutter" in Surbiton and went to Prague to confer with experts. A few weeks later, she wrote to D'Arcy again, this time from Oxford: "I have at last got my 'biological hypothesis.' I am so thrilled. Without your wonderful book I should *never* have got it....Enclosed is first draft of cytology paper...I *do* think morphology is electrostatic, don't you?"

Dot stayed in touch with her Viennese friends. They met again in Zurich at the International Congress of Mathematicians of 1932. In Vienna two years later, she spent an evening with Olga and the Godels and met Moritz Schlick, the center of the Circle, evidently for the first time: "What a ghastly wife—and he rather *too* sure of himself," she wrote in her diary. Then her Vienna vanished.

Olga Taussky read the swastika on the wall and left in 1934. After a year at Bryn Mawr College in Pennsylvania, she went to Girton on a Yarrow Fellowship; Dot met her at the docks and drove her to Cambridge. Olga's mother and sister escaped to New York; Dot stayed in touch with them to the end of their lives. Olga taught in London and married the Irish mathematician John Todd. After the war, the Todds emigrated to Pasadena and had long, successful careers in mathematics at Cal Tech. Despite Olga's ever-growing reputation, she remained a "research associate" until 1971.

On June 22, 1936, a deranged student shot and killed Moritz Schlick on the steps of the university. The assassin served two years of a ten-year sentence and then joined the Austrian Nazi Party. Karl Menger left Vienna in 1937. After a few years at the University of Notre Dame in Indiana, he moved to the Illinois Institute of Technology in Chicago. I met him in the early 1960s when I was a graduate student there. I'd had great teachers before, but never a course so gripping as his on "What Is a Curve?" When I read his posthumous memoir as background for this chapter, I learned a lesson in the transmission of scientific tradition: he'd taught me the very course his Professor Hahn had given, just as Hahn had taught it.[17]

I gave my first talk to the American Mathematical Society at its meeting in Berkeley in 1963. Just a "brief contribution," like Dot's talk on pleasures, but I was nervous and anxious as graduate students are, as she must have been. That evening, my thesis adviser took me along to a party high in the Berkeley hills overlooking the twinkling bay. He introduced me to the hosts, the celebrated mathematician pair Derrick and Emma Lehmer. He taught at

UC Berkeley; she worked at home. I also met the famous Julia Robinson, co-solver of Hilbert's 10th problem. Despite this coup, the University of California had not seen its way to hire her, but her mathematician husband was on the faculty.[18] I was as lonely at the party as I'd been at my talk, more so. At the talk, I only saw the chalkboard; here I saw my future. Then Olga Taussky Todd spotted me across the room, made her way over, and introduced herself, all smiles. "It's so nice to see another woman here!" she said. "Welcome to mathematics!"

In November 1933, Ada Minnie Souter Wrinch, her daughters' touchstone, was hit and killed by a bus. Dot rushed home from Paris. Her parents had been planning their first visit to Muriel and their five grandchildren in South Africa. Hugh decided to go alone, but couldn't bring himself to do it. He died a few months later, of a broken heart, his granddaughters say. Dot packed up her mother's two fur coats and the family silver and sent them to Muriel. The children used the silver in the sandbox. "My grandparents were way ahead of their time," says Talitha. "They were strong advocates of women's rights and sent both their daughters to university."

Alice Maud Procter retired as headmistress of Surbiton High School the same year Hugh and Ada Wrinch died. The three had lived to see a revolution in women's lives that they, in their own quiet ways, had helped bring about. Books and magazines urged girls to reject tradition and chart their own lives, adventures, and careers.[19] Surbiton Old Girls became nurses, doctors, dentists, radiologists, masseuses, missionaries, civil servants, teachers of everything "from arithmetic to ice skating,"[20] social workers, designers, illustrators, composers, concert artists, actresses, journalists, novelists and playwrights, beekeepers, gardeners, dairy workers, hairdressers, photographers, businesswomen, lawyers, leaders of women's institutes and organizations, a famous cinematographer, and even a member of Parliament.

And then there was the irrepressible Dorothy Wrinch. Miss Procter's ban on marks and prizes notwithstanding, the *Jubilee Book* listed Dot's endless laurels with pride.

When Alice Procter died in 1946, her sister Zoë received letters from Old Girls all over the world. "There was one common note in all these letters," wrote Zoë, "gratitude for the guidance and training for service to the world, which all had received from her."

"It was impossible to return to Paris immediately," Dot wrote to Lady Margaret Hall.[21] Instead, she plunged headlong into biology, chemistry, and botany. "At the John Innes I was occupied with gathering the data needed to test my theories as to the applicability of potential theory to chromosome mechanics. At Bedford, owing to the kind help and cooperation of Professor Wilson, I had the opportunity of testing various conclusions as to the influence of electrical factors to which I have been led during the last few years....My urgent need is to study certain problems of electrodynamics."

Brute facts were emerging from laboratories: experiments on the influence of electrostatic fields on plant growth, electroconductivity in sea urchin cells. Dot submitted a bold plan to the Leverhulme Trust for research in mathematics and cytology.[22] "My idea," she wrote, "is to discover and make precise what postulates as to the forces involved are required, if the known biological facts are to be deducible." Once she'd got the logical structure of plant growth and sea urchin cell division right, she would look for a law that predicted these special cases, a law describing "the mechanism by which genetic changes in the cell involve morphological changes in the organism."

Dot's Leverhulme proposal shows, more clearly than any other document of that time, how her "biological hypothesis" rested on, and drew on, her ideas of scientific method and the mathematics she'd been doing for the past 15 years. She didn't ask what genes *were*, she asked what they *did*. Each "genetic character," whatever the phrase might mean, corresponded to a specific potential field in the cell. The question was, how do these characters combine? Or, as she put it in her proposal, "Given a system of conductors whose capacities are known, how does the capacity of the system vary with the relative positions of the conductors?"

After solving that problem she would apply it, mapping the changes in the organism as she moved the conductors around, to "correlate the properties of the cell potentials with certain characters of the organism, and investigate in a manner mathematically precise the mechanism by which this correlation is effected."

"In the course of this work the problems of the arrangement of genes on chromosomes, in particular the cases of ring chromosomes, present themselves," she noted in passing.

But all this was preliminary. Dot aimed much higher: she would truss cytology with axioms, definitions, propositions, theorems, *Euclid/Principia* style. Then, and only then, morphology would be a mature science.

Dot worried, not that her proposal was impossibly ambitious, but that *she* might seem so. "If you could spare the time COULD you criticise this," she asked her friend Joseph Needham, in a cover letter to a draft.

I so much want to put the scheme in a form in which it will commend itself to the people who have to deal with it and I know that my standpoint is so different from the conventional biological standpoint that I have probably put it, all unwittingly, in a form in which my application may be unnecessarily prejudiced, either from the scheme being totally unintelligible or unnec. aggressive and self laudatory or what not.

You I think do understand the awful passions by which I am driven into these fields, but that will not be clear to them and is perhaps not an adequate excuse for my presumption....Oh dear. Do ask Dophy [Dorothy Moyle Needham, her friend and his wife] to help me cos she does I think understand that it is not evil pride that makes me think my life work is here, but just a drive from the subconscious that I can't resist.

The trust turned her down, but Dot was undaunted. She drafted plans for a book, *Mitosis: a Mathematical and Physico-chemical Study.* "I propose to publish this, by hook or by crook," she told her friend J. E. Crowther, *The Manchester Guardian*'s science journalist.[23] Might Oxford University Press be interested? Crowther, a longtime OUP editor, thought it would be, but the book project drops from the record. The problem of the arrangement of genes on chromosomes had "presented itself" to her, and she turned her attention to that.

Meanwhile, the Rockefeller Foundation was going through a metamorphosis of its own. In 1930, Dot's refusal to specialize was a fault; in 1935, it was a virtue, and just the virtue they were looking for.

The change, long discussed internally at the foundation, became evident in 1933 when Warren Weaver, the new director of the Natural Sciences division, emerged as a leader. Weaver, a mathematician, had been recruited by the foundation's new president, Max Mason, his former colleague at the University of Wisconsin and coauthor of *The Electromagnetic Field.* Mason and Weaver saw eye to eye. Instead of grants "to men of unusual promise in their respective fields," whatever those fields may be, they would fund mathematicians, physicists, and chemists to tackle biology.[24]

12 THE GIRL WHO KICKED THE HORNETS' NEST

"It was also hard work," Warren Weaver said of his ceaseless visits, in the first ten years of his reign, to laboratories and universities in England, Scandinavia, Germany, Switzerland, France, Scotland, Ireland, Holland, Belgium, Italy, Finland, Poland, the Baltic states, Austria, Czechoslovakia, Hungary, and southeastern Europe. He added the Americas, Asia, and Africa in his second decade. "We took many notes and wrote extensive diary every day—or, I should say, every night. One set of such visits did little more than furnish the basis for a continuing series of visits."[1]

Weaver, Tisdale, and other foundation officers asked advice from scientists of note and interviewed a vast pool of prospective grantees. Like nineteenth-century credit-rating investigators, they recorded a vast range of opinions, the vague and the tart and their own. There is only one way to judge a proposal, Weaver later explained.[2] You judge it by the capacity of the person to carry it out—his imagination, track record, expertise. Not by the proposal itself: good scientists can write bad proposals and bad scientists good ones. Not by letters of recommendation; they are never sufficiently candid. It's essential to get "indirect judgments about a man." You can't let the teller know you are considering the person for something. Your questioning should be indirect and in advance of the event. Who are the promising youngsters, and why? What are they doing? Why do you find it interesting? You don't ask, "Is young Dr. so and so good?"

This takes hours, Weaver continued. They have to believe in the importance of the Rockefeller Foundation and have confidence in you. "They know you're going to write that down in your diary but they know damned well that nobody outside of the Rockefeller Foundation is going to see the record.... You repeat this procedure time after time, place after place, and the network begins to close in. It almost automatically convinces that it will turn up good people…the pushy and spectacular, the dramatic individual, the awfully queer, unsocial, introverted individual." (But, Weaver did not add, if your informants share a bias, this process will almost automatically exclude.)

Weaver's net bagged brilliant characters with outsized personalities, voracious appetites, kaleidoscopic interests, irritating eccentricities, and hidden personal agendas. The kind who rush in where angels fear to tread. They didn't believe in angels.

Joseph Needham will leave his laboratory for China during the war and never return to it; he is remembered as "the man who loved China." At first single-handed, then with an ever-growing team of young historians, he roamed that vast country, uncovering the hidden history of ancient Chinese science and technology and recording it in 20 volumes. *Science and Civilisation in China* is a monument to Needham's vast scholarship and his inspirational gifts. But that came later. In 1932, he was a biochemist in the Cambridge laboratory of his mentor, F. G. Hopkins, and the author of *Chemical Embryology*. He was also known for his left-wing politics, idiosyncratic Christianity, encyclopedic grasp of the history of sciences, and expertise in Morris dancing.

Conrad Hal Waddington, "Wad" to almost everyone, was a study in Brownian motion. His Cambridge undergraduate career ran the gamut from geology to poetry to paleontology to philosophy and competitive running. In 1929, he wrote a paper on genetics, in 1931, a paper on mathematics; in between, he began research in embryology, collaborating with Needham, whose politics and Morris-dancing enthusiasms he also shared. Waddington would become one of the greatest developmental biologists of the century. In 1932, he held a research fellowship at the Strangeways Laboratory.

Rockefeller officers thought Waddington, Needham, and Wrinch, all three, were "queer fish." And who but queer fish could navigate the deep, swirling, hypercompetitive waters of Oxbridge and emerge with polymathic enthusiasms intact?

John Desmond Bernal was another. Dot called him "Desmond," but most friends called him "Sage." Sage for sagacity, not for the spice, though that too would have fit him. No one could match his encyclopedic knowledge of anything and everything from antiquity to tomorrow, his strobe-like flashes of insight into every scientific problem. He plowed new fields but left the sowing and prize-gathering to others; he was always onto something new, always somewhere else. But two signature traits kept grantors at bay: his prodigious womanizing and his devotion to the Communist Party. In 1932, Bernal was working, not happily, as an x-ray crystallographer at the Cavendish laboratory. His discovery that would turn biology inside out lay two years in the future.

In the early 1930s, Bernal helped revive the near-moribund Association of Scientific Workers that Dot and Harold Jeffreys had helped start 15 years before. She assisted him to some extent, but to what extent neither his nor her papers reveal. Nor have I found clues to her political views in those days of looming fascism. Suffice it to say, I suppose, that her friends—the Russells, the Needhams, the Crowthers, the Hogbens, the Bernals—were outspokenly left-wing.[3]

After Vienna, Delta (as Wrinch now signed herself), Bernal, Needham, Waddington, and Woodger, the biologician, began talking. The Theoretical Biology Club, which grew out of their discussions, is cited in histories of molecular biology for its pioneering vision. But they aimed higher: to explain the forms of living things.

D'Arcy Thompson had posed the challenge in *On Growth and Form*: "The biologist, as well as the philosopher, learns to recognize that the whole is not merely the sum of its parts. It is this, and much more than this. For it is not a bundle of parts but an organisation of parts, of parts in their mutual arrangement…this is no merely metaphysical conception, but is in biology the fundamental truth…" But what is "organisation"? Where does it come from? In the answers, this small group believed, lay the secret of life.

It was, they agreed, past time for that endless, ancient debate, vitalism versus mechanism, to go the way of the ether. They didn't agree on much else. The puzzle of growth and form will be solved by biochemistry, Needham insisted. The biochemist has *nothing* to offer, Woodger retorted. "It may be the case that to regard the traditional 'physico-chemical' ways of thinking as indispensable and as 'natural science par excellence' is the apotheosis of biological stupidity."[4] He drew up a truth table to show Needham his logical errors. What biology needs, said Woodger, is a Russell-style *Principia Biologica*.

The Woodgers lived with their four children in a large house in Epsom Downs, not far from Surbiton in Surrey. The spacious, shady lawn could be the setting for a long discussion of this and like matters on an August weekend. Woodger and Needham kept the group small, just their wives and Delta, Bernal, and Waddington. Eden Woodger had studied medicine. Dorothy Needham was a biochemist like her husband. Later she would be acclaimed for her research on muscle and one of the first women elected to the Royal Society.

In the picture of the group, Waddington smokes a pipe, on the left. Already balding at 27, he's the youngest in the group. Delta, 38 years old, stands to his left. Next to her is Bernal, 31. Then, continuing clockwise, we see Needham, 32, and his wife, 36. Woodger, 38, scowls between her and Wad. Eden probably took the picture. I don't know how old she was.

The host and prime mover led off on Sunday morning. Needham took sketchy notes. We are gathered, said Woodger, "to try to discern the form of the facts and frame the concepts in accordance with it."

The founders of the Theoretical Biology Club, undated, but probably taken by Eden Woodger at the group's first meeting in 1932. Clockwise, from the left: Conrad Waddington, Dorothy Wrinch, J. D. Bernal, Joseph Needham, Dorothy Needham, Joseph Woodger. Photograph courtesy of Gary Werskey.

What do we talk about when we talk about biology? First, we must use the right language. Not words like cells, organs, limbs; those are *instances*. Living creatures are organized, said Woodger, which means we should talk about hierarchies, levels, axioms, time, spatial association, relational maps between entities (in each of embryology, physiology, morphology, genetics).

But Waddington, next up, used the old language: flies' bristles, wing development, cell division. "W. thinks facts wrong," scribbled Needham. "W." must be Wrinch, not Woodger. "Is organiser all or nothing?...Organiser/organisee too simple...perhaps organisation centres have an infl. on one another."

In her own lecture, Delta played the role of math tutor, outlining the ABCs of topology and the theory of packing, the theory of sets, and how axiom systems work.

Bernal spoke last, unfurling a chart that put it all together. On a ladder labeled "hierarchy," Bernal named the rungs from the bottom up, in order of increasing biological complexity: "quantum mechanics, atoms (nuclei, whole atoms), molecules, crystals, colloids, polymer molecules, filter-passers, bacteria, nucleated cells, and then protozoa and metazoa (at the same level)." A ladder with mammals, not geometry, at the top.

If Dophy and Enid gave talks, Needham didn't describe them nor did he keep notes of his own.

After the group dispersed, Needham complained to Woodger that Waddington and Delta were planning to work "at the matter" together. He didn't approve; Woodger was the leader. As a grain of sand bestirs an oyster, the Theoretical Biology Club grew from Needham's irritation. "Why could we not get together again," he suggested, "the whole 5 of us (I except Dophi [his wife] and Bernal as not being essentially concerned) and sit down with a set of Normaltafeln [tables of morphological pictures showing normal development] and perhaps a set of standard models of amphibian development, and see what would happen. If you think of this, it seems to me that each would bring a valuable field of experience otherwise unattainable to the job, and that you being the leader of the group, we should get much further than any of us are likely to working in isolation."[5]

"Charmed to be a member of the Team," Delta replied from Zürich.[6] "Could turn up in London easily on Fridays or Sat.—less easily on Thursdays…NB. I expect I mentioned to you that I have worked out my ideas on cell division and that James Gray is at present vetting them. If he approves I shall publish them shortly."[7]

The club met as such for the first time in January 1933. Bernal and Dophy stayed in and Max Black, a logician, came too. Delta taught them the elements of Menger's dimension theory: "Lines of genes, strings of cells—1 dimensional array. Layers—2, Lumps—3."

But, Woodger insisted, a gene is not a *thing*, it's a concept, a "logical construct from Mendelian ratios."[8] He described it *Principia* style:

$$\text{mend} = {}_{Dt}\text{Aeq'xy}(x,y \; \varepsilon \; \text{whz. Apr'Zyg'x}|K|\text{Apr'Zyg'y} \; \varepsilon \; 1 \rightarrow 1)$$

Except in wartime, the club met more or less annually until 1952, but the membership changed gradually as the founders went their separate ways.[9] Delta stopped coming in the mid-1930s. Only Woodger stayed with it, rejecting genes to the end. A year after the club's last meeting, Watson and Crick announced the double helix.

The philosopher Karl Popper attended twice in the late 1930s. The club was "one of the most interesting study circles in the field of the philosophy of science," he said, and Woodger one of the most influential philosophers of biology.[10]

"We are not students of some subject matter"—Popper again—"but students of problems, and problems may cut right across the borders of any subject matter or discipline."[11] I agree.

I grew up with a Great Problem: my father spent 40 years on one that cuts right across the borders of biology, chemistry, criminology, medicine, neurology, psychology, sociology, and jazz. His laboratory was the Addiction Research Center, buried deep in a federal prison/hospital on 1,000 acres of Kentucky bluegrass.* We lived on the grounds.

I stumbled on another Great Problem my first year in college: *is there life on Mars?* A panoply of Nobelists filled the stage to confront the question. I remember, or I think I do, Harold Urey, the chemist, and Herman Muller, the geneticist. I know I remember the symposium's organizer, a graduate student in astronomy named Carl Sagan. In their hands, life on Mars was not about canals and little green men, it was chemistry, genetics, astronomy, geology, spectroscopy... Something clicked in me then: without fiction, we might not care; with only fiction, we might never know.

But, as I learned later when I began crossing borders instead of watching others do it, "interdisciplinary" and "cross-disciplinary" are easier said than done. Those borders are mined. Disciplining is what disciplines do: they train their adepts to ask certain questions and not others, to accept certain answers and not others. And disciplines, as Delta explained in her failed Rockefeller social studies proposal, control "the attitude of established practitioners toward free access to the profession of those with the necessary intelligence and training."

The Theoretical Biology Club took on a very Great Problem: "What is the relation between those large particles which we call elephants, trees, or men, and those extremely small ones which we call molecules or electrons?"[12] The founders—a bio-logician, an embryologist, a mathematician, a crystallographer, and a geneticist—had nimble, crosscutting minds. Yet they were not immune to misunderstandings. Confusion is endemic to all such discussions, whether between the sciences and the arts, between sciences, or even between science subspecialties.

What do we talk about when we talk about X? Biology is the science of structure, hierarchy, and organization, Woodger insisted, and everyone agreed. They didn't notice, at first, that they meant different thing by these words.

To Bernal, the words on his chart, from atoms to protozoa, were names for material objects. Visible, microscopic, or submicroscopic, the objects were *out there*. Not so to Woodger: those same words were instances of logical notions, just as $2 + 3 = 5$ is an instance of $x + y = z$. To Bernal, structure, hierarchy, and organization meant cells, tissues, organs; to Woodger, they meant symbols and Venn diagrams. Delta agreed with them both.

---

*The ARC grew into NIDA, the National Institute on Drug Abuse in Bethesda, Maryland.

What do we do about it? Bernal promoted the molecular vision of life. To understand how protein chains are arranged in the cell, you must understand their chemistry, their geometry. He convinced Delta to narrow her focus, from cells to chromosomes to proteins. But she kept her eye on the elephant. "I was awfully pleased with all the stuff in lec. 3," she wrote to Needham—he had asked her to read a set of lectures he planned to give at Yale—"cos it agrees so well with my impression as to the relevance of protein chemistry to the problems of morphology."[13]

To talk about morphogenesis, how form begins, Needham said, we must look into the cell and make cytology a molecular science. He didn't climb down Bernal's ladder to the protein molecule level. Instead, he asked how heredity was expressed. He put his hopes in True Analogy: just as magnetic fields rule physics, morphogenetic fields might rule biology.

Waddington used metaphor. Developmental decision-making was like a railway sorting yard, he said, "You are looking down an incline called the Hump. The wagons are pushed over the Hump and go running downhill and are sorted out by the systems of points into the various sidings. Now an embryo is in some ways analogous to a set of trucks sliding down the Hump. The first point is the primary organization centre and shunts off one set of trucks to the left, to become skin, and another set to the height to become neural plate."[14]

As Waddington developed his "epigenetic landscape," the railway sorting yard became a surface "down which a ball rolls. At various points, there are branches in the possible path the ball can take."[15] He didn't doubt that genes were material, but he thought they played supporting roles. Literally, supporting: "genes are visualized as pegs stuck in the ground, each with a guy rope attached to a sheet of fabric, which makes up the landscape. The idea is that individual genes have quantitative effects and their individual actions cooperate to produce the landscape as a whole."[16]

Delta, like D'Arcy Thompson, awaited biology's Newton. Woodger awaited its Russell. Secretly, they hoped to play those roles themselves. Meanwhile, they argued. Biology will never be physics, said Woodger. There will be no grand unifying theories. He told Needham he'd convinced Delta of that, but he hadn't and never would.

Old wine in new bottles. The spark of life: what is it? Rabbi Loew in medieval Prague brought the Golem to life by whispering The Name in its ear. Frankenstein bestirred his monster with a zap of electricity. In his Yale lectures, dedicated to the Theoretical Biology Club and published as *Order and Life*, Needham scoffed at vitalism. The rungs of Bernal's ladder were connected and biochemistry was the link. Thirty years later, in a new introduction to the lectures he wrote, "What I then affirmed,…may still be affirmed now, namely that biochemistry and morphology are very shortly going to blend into each other 'without any difference or inequality,' all barriers between them being broken down…If I thought like this in 1935, I am even more convinced of it in 1967; the bridge is almost built."

In his presidential address to the BAAS in 1933, James Gray called that affirmation a leap of faith. Delta too worried that Needham was masking a fundamental problem. "I can't seem to see eye to eye with you on one issue which I fear is an important one," she told him. "My feeling about Vitalism was always just this, why postulate some entity over and beyond those in the physical sciences, see whether you can't manage without…Well now, with regard to your organising relations why can't one say just the same?…I don't see that it is so much more respectable to allow in relations which don't belong to physics than to allow in entities which don't belong to physics."

In the 1980s, Francis Crick tried climbing Bernal's ladder and found he couldn't do it. "In spite of all their [his colleagues'] efforts, it proved impossible to get a foothold in the biochemical basis of the problem…and eventually I moved on to other aspects of the subject."[17]

Delta stayed with proteins, but always with the ladder in mind. From the protein's inner logic, all else would follow: not only the higher levels of biological organization, but also biology's higher levels of logical organization. Biology was, *sans doute*, on its way toward geometry.

In 1934, after two years of club discussions, two members huddled to work "at the matter" together, but this time they were Needham and Waddington. The matter was an institute of their own, a new cutting-edge institute of embryology. It should have its own building, not a stuffy attic at Strangeways or a tomb in the Cavendish basement. Their physicist-turned-author friend, C. P. Snow, got wind of their scheme and wrote his first novel, *The Search*. Delta got wind of it and reacted just as Needham had two years earlier, when she and Waddington were poised to veer off. Expand the discussion, she urged them. "I do feel most strongly that we really must put up a united front on this question. It is not in the least that I think that the cytology should be stressed, but rather that I feel that it is now or never if we are to claim a real Morphological Institute."[18] X-ray crystallography, the tool of the future, must be part of it too. They took her point. On the back of her letter Needham wrote, "State agreement with general line of P. C. Morphology.…Involves accomod. for cytology, crystal physics. Sketch out organization. Clear up dead issues. Our mind made up on site (*before too late*)."

In this scheme, Needham, Waddington, Delta, and Bernal would be salaried department heads: embryology, genetics, cytology, x-ray crystallography.

Research associates and assistants would work under them. Woodger would be a regular guest. The institute would be clubby, discussions spontaneous: they wouldn't have to dash across town. They would have steady incomes at last.

Waddington needed the money for alimony for the wife he was divorcing. Bernal was poorly paid and underrecognized. Delta, with a child to raise and a career to forge, needed the institute at least as much as they did. Her not-officially-ex husband was consigned to Warneford, her sister lived halfway around the world, and both her parents had died. Torn between research and motherhood, Delta looked for a boarding school for Pamela. But, she found, "only those schools which have some superstition to conserve and instill into the young have funds…eg also the Friends' schools. e.g. again coeducation runs contrary to many vested ideas—thus there are very few funds for coed education." St. Christopher's in Letchworth turned out to be suitable. Still, Delta felt trapped in Oxford, tutoring in all five women's colleges to make ends meet. The institute's salary would cover Pamela's tuition and let Delta work in peace with colleagues she adored.

The Rockefeller Foundation would, they all hoped, pay for the institute and pay everyone in it. Friends put in good words. "I had the Rockefeller people—Weaver and Tisdale—here last week," wrote Francis Crew, the director of the Genetics Institute in Edinburgh that Waddington, which later head. Crew had fallen suddenly, madly in love with Delta, but she saw no future with him, and the relationship settled into friendship. "They eagerly accepted my views that Wrinch and Waddington were [?] people to be watched and encouraged and that Dorothy Wrinch ought to be removed permanently to Cambridge to be alongside Gray, Needham, Waddington and the rest. I gathered that something was afoot and that I was confirming their own decisions. I hope that I was being helpful. Certainly I was helping the development of genetical science—or trying to do so."

Something *was* afoot, but not what Delta expected, and not, I think, quite what she wanted.

"On the Molecular Structure of Chromosomes": the lecture, on May 13, 1935, at the University of Manchester, was Delta's post-chrysalis coming-out party. "To my friend S. M. F.," she wrote in the printed version.[19] This time the dedicatee is unmistakable: Sara Margery Fry. Delta thanked a battery of colleagues in genetics, chemistry, crystallography, and mathematics, but only Bernal from the club.[20]

"A number of new sciences have passed from the embryonic stage. Discarding description as their ultimate purpose, they are now ready to take their places in the world state of science." Delta didn't thank the Aristotelian

Society, but her debt is clear. "The thesis which I wish now to develop is but a logical consequence of the thorough going application of this principle. I visualize the animal and plant morphology of the future, and particularly cytology, as a member of the world state."

Potential still had potential: "Is there indeed anything fantastic in looking forward to a future in which the various entities in the cell-centrosomes, spindle, nuclear membrane and most fundamental of all the chromosome itself, are recognized as themselves providing the most delicate and illuminating objects for the demonstration of the principles of the mathematical theory of potential and the concepts of pure physical chemistry?"

But "the molecular aggregate which is the chromosome" was enough for one day. At length and in detail, Delta proposed a model of the chromosome as a molecular fabric: proteins laid down like a warp, nucleic acids threaded under and over like a woof. The two ends of each acid, hanging from this hypothetical loom, linked up in the living cell, she said. The chromosome is a woven sheath.

And heredity? Where did Delta put the "genes"? "I locate the genetic identity of a chromosome in its characteristic protein pattern," said Delta. Genetic identity is expressed by the sequence of twinkling Christmas tree lights. "The genetic constitution of a chromosome is to reside not only in the nature of the [amino acid] residues of which it is composed, not only on the proportions of these different residues, but essentially and fundamentally in their linear arrangement in a sequence of sequences." The twinkling lights are a *code*. Your code. Your sequence is you.

"I have been putting together various statements on the sequence hypothesis to see how far back the idea goes and in what forms it has been expressed," Robert Olby wrote to Dorothy in an undated memo, probably 1970.[21] He was drafting *The Path to the Double Helix,* a comprehensive history of the discovery of the structure of DNA.[22] "I was particularly pleased to find what you wrote in *Protoplasma* in 1936. This would appear to be a refinement of the ideas you presented to the Theoretical Biology Club."

"It was not an obvious step to make the connexion between the linear sequence of the genes and that of the amino acids in a polypeptide chain," Olby wrote in his book (emphasis his). "One person in the thirties who did just this was Dorothy Wrinch, the British-trained South American mathematician [*sic*]. At an informal meeting of the Theoretical Biology Club in Cambridge in 1934, she expressed the view that the specificity of genes resides in the specificity of their amino acid sequences."

After the meeting in Manchester, Delta's host, Michael Polanyi (then a physical chemist, later a philosopher), gave her a rundown on his colleagues' reactions. That the chromosome must have a fiber structure is a good idea, as is connecting this with the genetic functions. The nucleic acid warp is sound but the protein woof is tentative (he'd interchanged warp and woof). "To attribute to the side chains the genetic identity of the chromosomes is also a brilliant idea, but one cannot feel just as sure that it is true....Speaking as a distant relative of the gypsies," he concluded, "I might add, as a piece of fortune-telling, that you will live to great recognition of your vision."

Delta's vision was ignored. "More might have been made of Wrinch's sequence hypothesis had not her molecular model of the chromosome come under attack," says Olby. The warp-woof rightangles implied optical properties for the chromosomes that were opposite those actually observed. Crick formulated the modern sequence hypothesis in the late 1950s. In the afterglare of the cyclol controversy, to which we turn next, Dot's papers on the subject were invisible.

But Warren Weaver noticed at the time. The Rockefeller Foundation had refused to fund the morphology institute. *Brains over buildings* was its in-house motto, and why stir up murky Cambridge politics? Let the university decide whether, where, and when to build it, and let them pay for it. As for the brains, Waddington's divorce was no concern of theirs. Funding Bernal might enrage powers they needed on their side. Even the shy, genial Needham had detractors. The foundation was cautious: their program was new, they couldn't take chances. Give him an assistant, maybe, but not everything he asked for.

Dr. Wrinch, on the other hand, was making remarkable progress along the very lines Weaver was hoping for. In September 1935, he presented the case to his board.

The journal *Nature* announced the Rockefeller Foundation's five-year grant to Oxford in November: "This grant is to enable Dr. Wrinch to continue and develop her researches into relationships between chromosomes and protein aggregates which have been the subject of several notable contributions...Her recent work on the structure and behaviour of chromosomes in relation to protein aggregates is a new field of inquiry from which further results of high importance may be confidently anticipated."[23]

In Leeds, England, Rockefeller officer W. E. Tisdale wrote in his diary:[24]

A. tells me we shall probably be criticized by biologists for the Dorothy Wrinch project more than for any other one we can do. Following on H's remarks this seems perhaps to be a correct prognosis.

IV PROTEINS AND THE IMAGINATION

The true test of an hypothesis, if it cannot be shown to conflict with known truths, is the number of facts that it correlates and explains.

—FRANCIS M. CORNFORD, *The Origins of Attic Comedy,* 1934

13 BUZZ

Hornets build their nests by chewing old wood—tree bark, rotting fence posts—and mashing it into a paste, says John Nicholson's ancient copy of *Insect Architecture*. They use it to build cells for themselves and a shell around the colony. They make a buzzing sound, like bees.

"We took many notes..." Rockefeller Foundation officers' diaries and letters convey the texture of the time. Their process did not exclude Delta, but neither did the network close in on her. I have selected but not condensed the extracts that follow, and present them verbatim, in chronological order. (And I have, for consistency's sake, substituted initials for names in the few cases that the officers did not. Usually, the informant was only identified in the entry heading; initials were used in the body of the note.) To learn who's who, see the endnotes.[1]

1935

Pasadena. B...is very keen for Dorothy Wrinch's work.

London. Mathematical and theoretical people do not have a ghost of a chance to contribute anything to theoretical biology. H. feels that a properly trained chemist or physicist (and he leans towards a bio-physicist)—is probably the type of individual who will develop a theoretical biology, if one is developed.

1936

London. v. M., without knowing our interest in D. W., recommended that I should read her recent article in "Nature," in which she presented some of her studies in protein structure. He expressed the opinion that these studies of D. W. represented what was probably the most significant progress in protein studies within the last 8 or 10 years...His observations are predicated on the belief that one equipped to deal with geometrical conceptions would do what chemists apparently had not been able to do.

Others admit to not reading her paper because of prejudice.

Leiden. HMM asked G. what he thought of W.'s recent attempts. His reply was that she perhaps goes too far and then publishes, whereas he (G.) goes too far and does not publish. However he added quickly that he did not really mean to criticise W whose work is most stimulating.

London. K's opinion of DW and in fact also of Dorothy Jordan Lloyd is not very high. L. expressed the opinion that Dorothy Wrinch is in bad favour in many quarters in England…

Cambridge, UK. In leaving HMM inquired cautiously concerning Dorothy Wrinch's work on proteins. P. said flatly that he has always said she is a fool but that B. insists that she is only mad.

London. C. being a protein chemist the conversation turned on to Dorothy Wrinch. Both D. and C. expressed themselves, without at first knowing our aid to DW (?), to the general effect that everyone in England in or near the protein field is more than antagonistic to her.

H.* went to extreme lengths in damning everything DW has done. Such statements as the following were used by C., D. and H.: "W. has been engaged in plain poaching," and "everything she has published has been most obvious to most real protein chemists" and "she expects everyone to drop everything and to work for her."

C., who was the least caustic in his statements said that he has long since ceased to tell her anything of the work in progress in his laboratory, because he is certain that it would be published as DW's idea in a note to "Nature." All three agreed that they could not understand how such a good protein chemist as Dorothy Jordan Lloyd had allowed herself to get into the position of published jointly with DW…

Although all of these statements and others were made in apparent honesty, it is evident that there exists some little prejudice against DW because she is a woman. The statement was made that she had recently submitted a paper to the Royal Society which was turned down. Although the author has no means of knowing who the referee is on a submitted paper, DW thinks that she does in this case and "she went about the country saying very nasty things about the wrong person." The group of three felt this to be a characteristic feminine reaction.

1937

Cambridge, MA. C.* is not at all sure that all of W's ideas are sound but nevertheless he feels that she is undoubtedly contributing in a significant way to the advance of knowledge on proteins. She is certainly active as a ferment, introducing new ideas and arousing new interests. C. is inclined to forgive her for some of her personal peculiarities partly on the grounds of her tragic personal life, and partly because he considers her, along with her colleague Dorothy Jordan Lloyd, typical instances of the Pankhurst feminist movement in England.

H.** is very much interested in W's work and feels convinced that many of her ideas are both sound and important.

Baltimore. J. is still talking about his most pleasant experience at Oxford. He became acquainted there with Dr. Wrinch, saw her again when she visited the Hopkins this past winter and received a most excellent impression of her as a person of unusually valuable ideas. He also has heard criticisms of Dr. Wrinch but seemed inclined to feel that the contributions she will probably make would seem to outweigh any minor deficiencies.

Vienna. In some general conversation with M., who has just returned from England and Holland, M. says of Dorothy Wrinch that at least 60 percent of her ideas are worthless or impracticable, but that the rest are of great value—unique and worth whatever they cost.

Geneva. Of Dorothy Wrinch, Me. (to paraphrase a once governor of Texas) damns her and her ideas from Dante's Inferno to Breakfast. It may not be amiss, however, to remember that Me., with M., claim to have developed application of X-ray studies to rubber and cellulose problems.

Cambridge, UK. Knowing P's low opinion of Dorothy Wrinch, HMM asked for it in the light of recent publications. P's only comment was that W. "has now completely sunk herself" by her last brief letter to Nature.

London. When Ha. mentioned Dorothy Wrinch's name, HMM asked him his present opinion of her work. He said that W. fitted her last theory to a protein number which Bergmann published and which may not be (probably is not, according to Ha.) correct. He stated emphatically, however, that W's efforts are worthwhile, will probably stimulate experimental work.

Chicago. Ho. re Dr. Dorothy Wrinch. Interest—the proteins and everything else. Does not have much knowledge of chemistry, but certainly is out to learn. Her speculations on protein structure will certainly stimulate further work. A good influence in this field. She is bent on getting the protein ball rolling. Overflowing with energy and enthusiasm. Energetic, a hard worker, entirely devoted to science.

Oxford. I am surprised to hear from R., Y., Z., F., P.*, all independently, of their real interest in her work as it affects theirs. She seems actually to be getting on in Oxford as well as she reports she did in the U.S.

Paris. Perhaps a similarity in B*'s ideas and those of E, who was in a few days ago, indicates a common leaven, the catalyst probably being Dorothy Wrinch.

New York. Just within last week I again had both sides of the story. I was calling on P.** at Ithaca...He was in bed with a cold, and a Commonwealth fellow from Oxford was also calling on him. When Dorothy Wrinch's name was mentioned he blushed furiously and had to draw on the deepest reserves of the English character to keep from being profane in Mrs. P**.'s presence concerning what the Oxford chemistry crowd thinks of her...On the other hand I was talking to U. ...and in commenting on a woman in his laboratory he said that she was exceedingly good but of course not a Dorothy Wrinch. When I asked him what he meant by that, he said: "Well, I mean of course she isn't an outstanding genius," going on to explain that he supposed there could be no serious difference of opinion that that is what Dorothy really is.

Paris. Et. deposed German now in the Cancer Institute of Lisbon, where he has the dept of physical chemistry. Et. is here for the Reunion International de Physique, Chimie et Biologique. He was particularly interested in the lecture of Dorothy Wrinch because of his past and present interest in proteins. He said that DW gave an excellent paper but that the discussion of it was terrible, in the sense that so few investigators knew enough to appreciate her viewpoint. When questioned concerning young protein chemists of distinction in Europe Et. said simply that they almost did not exist. He had no names to suggest. Some years ago he had published a paper on straight physical chemistry of proteins, and had stated that his findings could only be explained on the basis of great symmetry of the protein molecule. DW's recent work confirms this, according to Et.

1938

Woods Hole. S. is spending the summer at Woods Hole, writing papers. He also attended the Cold Spring harbor Symposium for a few days, before coming here. Regarding Dorothy Wrinch, he thinks she has good ideas but is much too "preachy" about them. But even if her theories are wrong, they have stimulated chemists in Europe and America to think about protein structure and they will continue to think about it and work on it regardless of what happens to W's theories. In this respect, support for her has been fully justified in S's opinion.

St. just back in Woods Hole after 10 days at the Cold Spring Harbor Symposium...Dorothy Wrinch was a storm center for about a week. The group, with the exception of Langmuir, was pretty largely opposed to her ideas, and Astbury was the leader of the opposition, being ably supported by Northrop of the RI and others. Wrinch's methods of argument in the discussion were tricky. When detailed facts were presented contrary to her theory, she declared she had no interest in such details but was looking only at the broad general picture, although a few minutes later she would bring forth the minutest detailed facts in support of her theory.

London. On D. Wrinch, Br. confesses that he has not studied her papers, and in the Langmuir-Wrinch paper of September 1938 in the Journal of the American Chemical Society hc has let Langmuir's name lead him to believe there must be something in it. He suggests that B.* has studied this paper and that a safe opinion would come from him. Says W. is such a nuisance that it is hard to take a sympathetic view of her work, although there is no question but that she has stimulated the subject—as well as a lot of animosity.

14 THE CYCLOL MODEL

Greeks saw Orion the hunter. Indians saw a deer, Australian aborigines a canoe.[1] "Different tribes projected different images," E. H. Gombrich observed in *Art and Illusion*, "and nothing is more instructive than to compare the different interpretations given to the same group of stars."[2]

Delta's protein model was a cultural Rorschach test too. What were the tribes, what did they see, and why?

Just as "crystal" once meant quartz, "protein"—from the Greek *prota*, "of primary importance"—was first used for a substance found in casein (*caseus*, Latin for cheese), fibrin (a blood-clotting factor), and egg white. Terminology proliferated as muscle, hair, collagen, keratin, enzymes, antibodies, and hemoglobin were found to be protein-like. In 1907, the International Congress of Physiologists cut through the confusion and declared them all proteins.

Early in the century, Emil Fischer (Nobel Prize in Chemistry, 1902) suggested that proteins are peptides strung out in chains, with amino acids dangling. A guess it remained. "The unravelling [of protein architecture] apparently lies years and years in the future, so far ahead that most protein chemists will not talk about it," an eminent chemist wrote in 1936.[3] This, says the *Oxford English Dictionary*, was just the second time the expression "protein chemist" appeared in the English-language literature.

In the 1920s and 1930s, Bill Astbury, a physicist at the Textile Physics Research Laboratory at the University of Leeds, used x-ray diffraction to show that muscle, hair, collagen, and keratin are fiberlike on the atomic scale. That's why silk is silky, why asbestos flakes, why crimped wool stretches in water. Astbury deduced that parallel wool protein fibers cross-link and the links can flex: up, crimped; down, stretched.

But as Delta delved deeper in the protein literature she began to doubt that the chain hypothesis could explain anything more complicated than flexing. Enzymes, antibodies, albumin, hemoglobin, insulin, the proteins essential for life: how could chains explain the different atomic weight classes, the beautiful crystals proteins form under careful coaxing, and their mysterious folding and unfolding? How could a tangle of tens of thousands of atoms save a dying child?

Peptides *might* form long chains, as Fischer had proposed and most chemists believed, but nothing suggested they *had* to. Delta wasn't the first to suspect they formed rings instead. "I happened to come across a reference in [illegible]'s *Biochemistry* to the work of Scadskow and Zelinsky who don't believe that the protein molecule is a lovely chain but rather a system of cyclic structures," she wrote to Bernal in 1934.[4] (Her sketch of their diagram is nothing like a cyclol.) "Is this of interest to you?" It hadn't interested Bernal or Delta then, but it interested them now. Rings would explain the small numbers of different molecular weights, and maybe the nature of denaturing.

As Delta puzzled over rings, Bernal told her something he'd heard through the grapevine. The exact lineage of this suggestion is repeated like a mantra in every discussion of the cyclol controversy, so here it is, for the record. Three years before the time we are speaking of, a young Oxford chemist, F. C. Frank, attended one of Astbury's lectures on wool. Someone in the audience asked the speaker why wool crimps: why, chemically, the cross-links flipped up and down. Astbury had no idea. Frank did: the links would flip if the hydrogen atoms left the nitrogens to join the carbons on their left, as in the before-and-after diagrams of figures 4 and 5: look closely. (Note: I use a dashed line to emphasize the transfer, not to indicate bond strength as a chemist would.)

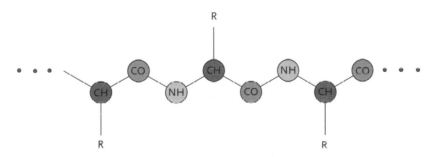

Figure 4 A schematic drawing of a fragment of a polypeptide chain.

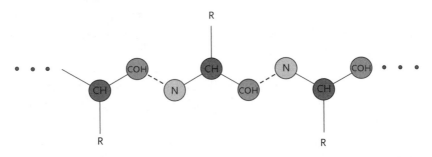

Figure 5 A schematic drawing of a fragment of a polypeptide chain, modified as suggested by F. C. Frank.

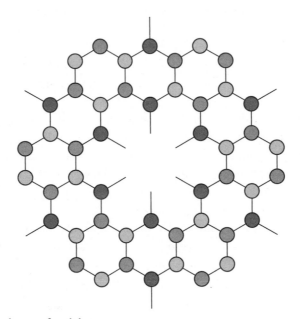

Figure 6 A cluster of cyclol rings.

Aha! said Delta. The nitrogen atoms, now untethered to hydrogens, could link to a second carbon and branch out from a linear chain into a two-dimensional pattern. The tetrahedral bonding of the C's and the N's would force the peptides into rings. She dubbed the rings cyclols. The name, if not the notion, caught on instantaneously. As she saw it, the linking would go on and on, rings joining together to form a molecular fabric. Not a loom-woven fabric with warp and woof like her now-discarded chromosome; the protein fabric would be elegant, intricate, lacelike (figure 6).

Delta described her vision of the cyclol fabric in *Nature* and Frank wrote a note in support. She and her fabric became the talk of science lounges,

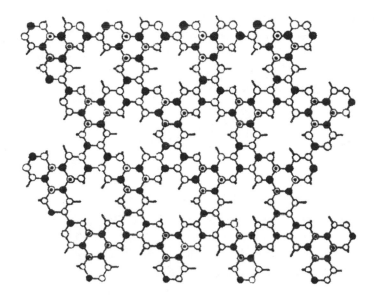

THE CYCLOL PATTERN. THE MEDIAN PLANE OF THE
LAMINA IS THE PLANE OF THE PAPER. THE LAMINA HAS
ITS 'FRONT' SURFACE ABOVE AND ITS 'BACK' SURFACE
BELOW THE PAPER.

● = N.

○ = C(OH), PEPTIDE HYDROXYL UPWARDS.

◉ = C(OH), PEPTIDE HYDROXYL DOWNWARDS.

○— = CHR, DIRECTION OF SIDE CHAIN INITIALLY
    OUTWARDS,

○- = CHR, DIRECTION OF SIDE CHAIN INITIALLY
    UPWARDS.

A fragment of Dorothy's cyclol fabric. From "On the Structure of Pepsin," by D. M. Wrinch, *Philosophical Magazine*, seventh series, 24 (July–Dec. 1937), 941.

lecture halls, symposia. "Whether or not this particular theory proves true," wrote H. T. Pledge in *Science since 1500*, "the day of geometry seems to be dawning in this subject."

Later Delta would be roundly, vehemently, and perhaps justifiably accused of clinging to her cyclol hypothesis at all costs and at great cost. But it should be noted that she had conceded the faults in her chromosome model, and she modified this first version of her protein model. The fabrics stacked in layers, she had said, but Bernal gave chemical reasons why they couldn't. Very well, then: she would make a three-dimensional model.

On one of his visits to Oxford, Delta showed Warren Weaver "various nets of hexagons, pentagons, etc. which can cover a sphere without a singular point. She has a somewhat vague but interesting idea for the development of a new (?) branch of mathematics which she wishes to call 'discrete topology.'"[5] He should have told her to go for it; it was an excellent idea, ahead of its time. Instead, he "tried to get her to be more definite, to see, for example, what relation there would be between this subject and group theory; but this attempt is not successful."

Weaver's diary entry expanded to a mini-essay. "W. is a queer fish, with a kaleidoscopic pattern of ideas, ever shifting and somewhat dizzying. She works, to a considerable extent, in the older English way, with heavy dependence on 'models' and intuitive ideas." Writing in the third person (as usual in his diaries), he continued, "On the basis of her former work in mathematics and physics WW had anticipated a more systematic analytical procedure; and finds it very hard to judge W. There is no question but that she has already entered deeply into the wide variety of topics that bear on her general problem of protein structure; and she has obviously impressed a number of critics."

Delta's "three-dimensional aggregates" crystallized quickly. As Weaver's note suggests, she tried folding her fabric like origami. Fold, fold, and fold again: curiously, the Shaker religious sect, long ago, had linked this symmetry to simplicity:

> When true simplicity is gain'd,
> To bow and to bend we shan't be asham'd,
> To turn, turn will be our delight,
> Till by turning, turning we come round right.

Since the bonding angles of carbon and nitrogen are tetrahedral, said Delta, the fabric must fold at the same angle as adjacent faces of a tetrahedron. Which polyhedra, besides tetrahedra, have precisely these angles and no others? Tetrahedra with their corners snipped off.* They come in different sizes. She called the smallest C1, the next C2, then C3, and so on. The series is simple and beautiful.

In the autumn of 1936, Delta sailed to America at her own expense to test her ideas on American audiences. Weaver contacted chemists, biologists, and physicists throughout the northeast. Dorothy Wrinch was coming for a two-month visit. Would they like to meet her? Yes, they replied, and would she give a talk? Delta came, saw, charmed, and infuriated. She catalyzed thought, debate, collaborations, and outrage.

---

*Casual readers thought she'd proved there were no other possibilities. But she usually made it clear that there might be more.

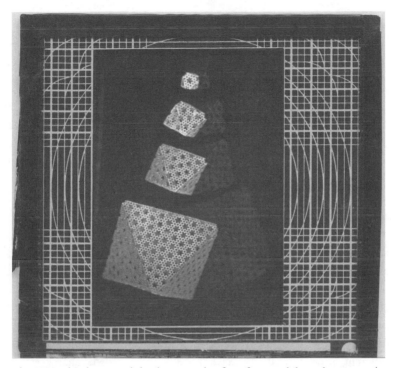

Dorothy Wrinch's lantern slide showing the first four cyclols in her series (top to bottom: C_1, C_2, C_3, C_4).

Ross Granville Harrison, the renowned embryologist at Yale, didn't let Weaver finish his question. Interrupting, he asked him "if he knows anything about W., and whether she would be approachable, interested, and available to come to work with him for a year."[6]

Harrison had been voted the 1917 Nobel Prize in Physiology or Medicine for his development of the technique of tissue culture, but the foundation did not award it to anyone that year. Now, in 1936, he studied asymmetry in the developing embryo. Protein molecules were, he thought, the key to it. "Harrison speculated that primary polarization of egg arose from alignment of the protein molecules," explains Donna Haraway in *Crystals, Fabrics, and Fields,* a history of embryology in the 1930s.[7] "The fundamental organizing insight in Harrison's approach to development was his conception of the role of protein molecules."

The details of Harrison's meeting with Delta are not recorded, but they must have discussed Astbury's work on fibrous proteins, since Harrison soon went to see him. "I am delighted to learn that you are coming to the states again," he wrote to her a year later, "and hope that you can spend at least several days in New Haven, since I need your help very much."

Irving Langmuir hadn't thought about proteins until Delta visited the General Electric Laboratory in Schenectady. He had thought about everything else: "so steady was his output and so high the level of its excellence," says an adoring biographer, "that one can, at best, imitate the biographers of artists—and write, say, of Langmuir's Surface Chemistry Period, as one would of Picasso's Blue Period."[8] Nineteen thirty-six was that period's Thin Film subphase. Langmuir created layers one molecule thick: oxygen on tungsten, oil on water. A relatively small number of layers produced optical effects from which he could measure the dimensions of the molecules themselves, and the absolute values of x-ray wavelengths.

"Until recently, however, we have not seen how these results could be directly applied to human problems or to human needs," said Langmuir.

Last December Dr. Dorothy Wrinch visited our laboratories and asked if we could apply the technique which we had developed in the study of multilayers of fatty substances to the study of proteins. After a few days of work, we found that we were able to build up multilayers of any protein on metal plates and in that way we were able to study many new properties of these important substances....Each different protein that we have tried gives its own characteristic thickness by a measurement which takes only a few minutes. ...This method of rendering these thin films visible should have great value as a biological tool; very likely it will find a place in the diagnosis of disease. It has the advantage that it makes possible the detection of extremely minute amounts of substances from very dilute solutions...Thus these experiments with multilayers seem to open up several promising lines of research which may well have applications to human needs.[9]

In New York City, Delta met with Harold Urey at Columbia University and Harry Sobotka at Mt. Sinai Hospital. Urey had won the Nobel Prize in Chemistry in 1934 for his discovery of deuterium, a heavy isotope of hydrogen. Because isotopes can be tagged and traced in living plants and animals, he thought they might be used for medical diagnoses. In 1936, he was studying heavy oxygen. Harry Sobotka, a versatile biochemist, was a cultured Viennese. With Delta as muse, Urey and Sobotka began a research project on isotopic interchange in proteins. Sobotka "has tentatively promised W. to attempt to make some compounds for her and W. is herself prepared to go on for some considerable time with theoretical considerations, but realizes that the organic chemists will not remain interested in her ideas very long if she does not produce something 'in a bottle.' She continues insatiable in her ambition to have everyone work on her ideas," Weaver noted.[10]

Delta engineered a collaboration between Bill Astbury and the geneticist C. W. Metz at Columbia University on the x-ray diffraction of chromosomes, and another between Astbury and Harrison on embryology. She sparked a project directed by Robert Robinson at Oxford and galvanized insulin research in Niels Bohr's laboratory in Copenhagen. The Rockefeller Foundation funded them all.

The foundation had been nudging Niels Bohr toward biology since Weaver came on board. Bohr hesitated at first, then waxed enthusiastic.[11] By 1937, his research group was tracing tagged atoms in metabolism and circulation.

Niels Bohr had absorbed physiology by osmosis at the family dinner table. He still sympathized with his father's rejection of Man as Machine. When John Scott Haldane argued, in that 1918 debate with D'Arcy Thompson, that simplicity dissolves to complexity close up, he spoke for his friend, the elder Bohr, too. "On the one hand," Niels Bohr told a conference on light therapy in Copenhagen in 1932, "the wonderful features which are constantly revealed in physiological investigations and which differ so remarkably from what is known of inorganic matter have led biologists to the belief that no proper understanding of the essential aspects of life is possible in purely physical terms. On the other hand, the view known as vitalism can hardly be given an unambiguous expression by the assumption that a peculiar vital force, unknown to physics, governs all organic life."

Bohr thought he saw a way out of this maze: life is like the uncertainty principle of quantum mechanics. Just as the position and momentum of an electron can't be measured (with high accuracy) at the same time, "we should doubtless kill an animal if we tried to carry the investigation of its organs so far that we could tell the part played by the simple atoms in vital functions. In every experiment on living organisms there must remain some uncertainty as regards the physical conditions to which they are subjected, and the idea suggests itself that the minimal freedom we must allow the organism will be just large enough to permit it, so to say, to hide its ultimate secrets from us."

Bohr's lecture, published as "Light and Life," inspired the young Max Delbrück to leave physics for biology, but Joseph Needham scoffed. Bohr is "notably naïve" about what death is, he snapped in his lectures at Yale. Death is not one single thing: "there is one death when the metazoan body ceases to perform its normal functions, there is another death when the tissue-slice isolated from it ceases to glycolyse or to respire in the manometer, there is a third death when the cell-free enzyme preparation isolated from the tissue-slice ceases to catalyse its appropriate reaction."[12]

The Copenhagen isotope project suggests that Bohr took Needham's point, up to a point: life *could* be studied, even on some microlevels, without killing the living animal. But still he puzzled over the great either-or. A letter Delta wrote to him in 1938 suggests Bohr thought cyclols might hold a clue. "I have been thinking a great deal about what you said to me in Birmingham and earlier in Copenhagen," she wrote. "I really think that there may be something about the cage structure which does represent the first step from the inorganic to the organic. Evidently in the phenomenon of death there is no instantaneous change of chemical composition. Nevertheless, there is an over-whelming change which in fact does occur. This, not being a change in chemical composition, can only be a change in structure and this is, I suppose, where the intrinsic importance of cages comes in. I wonder if I am interpreting rightly the things which you said to me. The point of view is growing on me daily and also on Dr. Langmuir."[13]

Asymmetry in the embryo, diagnostic tools for medicine, oxygen exchange, life and death: the cyclol catalyzed the scientific imagination. "As soon as the chemist discovers exactly how the protein molecule is constructed, living processes will no longer be mysterious," said the *New York Times*.[14] Cancer and sex hormones will be understood if she's right, Irving Langmuir predicted. Albert Szent-Györgyi, the 1937 Nobel Prize in Medicine or Physiology laureate, waxed even more visionary. Said the *Times*, "He pictures the physician of the future telling the chemist just what effect he wants to produce in a sick patient. Whereupon the chemist will sit down, draw a picture of the wanted molecule, synthesize it and send it on. Prescriptions will become problems in geometry. A tall order, Szent-Györgyi admits, yet one that we ought to be thinking about now."

"Without doubt there is a great deal of truth in her general picture," Linus Pauling told Weaver in March 1937.[15] "This picture is, however, still very far from definite—she suggests various alternatives and does not make any definite predictions." (He had not read her recent papers, nor had he been following the story closely: all the collaborations she fostered were designed, in part, to test specific predictions.) *The Nature of the Chemical Bond* had not been published yet, but Pauling was already the reigning authority on the subject. "I have felt," he continued, "that the definite suggestion which she did make regarding protein structure, dealing with a type of polypeptide condensation involving hexagonal rings, is incorrect, since Mirsky and I came to the conclusion from the consideration of available experimental facts that the structure of native proteins is determined by hydrogen bond linkages," not a special cyclol bond as she claimed. Pauling's criticisms would grow harsher, and then still harsher.

Other chemists found details troubling too. "This particular model by Dr. Wrinch cannot be correct," Maurice Huggins of Eastman Kodak told the American Chemical Society.[16] "It requires neighboring atoms to overlap their centers, being much closer together than is known, from X-ray studies of small molecules, to be possible." Huggins proposed a variant resembling a doughnut, also hexagonal, but his rings were held together with hydrogen bonds. This will, of course, have to be tested, but "it can truthfully be said that the prospect of learning the details of the structures of complex protein molecules absolutely essential to an understanding of life processes is now much brighter than it was a few days ago."

"The cyclol theory partly grew out of conversations with Bernal and myself," Dorothy Hodgkin wrote to me after Delta's death. "We were both convinced it was wrong almost immediately in 1936 as we read up more protein chemistry ourselves and did experiments. But our arguments, of a chemical kind, on the sort of residue packing with the cyclol molecules, the hydrolysis fragments of protein and Astbury's fibre patterns in relation to globular proteins, did not convince her. They did not convince Langmuir either."

Nothing is more instructive than to compare the different interpretations given to the same group of stars.... Reflecting on his long career chronicling scientists, J. G. Crowther wondered why it was that in the grand debates over the nature of light, x-rays, and cathode rays, the British opted for particles and the continentals for waves. Perhaps because the Brits played cricket while the continentals played music?[17]

Who were the cyclol's opponents, who its defenders? Geography did not divide them. Nor was it a battle of experimentalists *versus* theorists: there were some of each on both sides, and Langmuir and Pauling were something of both. Nor was it a battle of young *versus* old or men *versus* (the very few) women.

Disciplines, perhaps? Dorothy Hodgkin described it to me that way. "The cyclol theory as proposed had certain attractive features which were welcomed particularly by physicists and biologists. By physicists, because it provided a reason for the existence of large molecules of definite sizes, by biologists because it provided specific patterned surfaces thought to be desirable for enzyme reactions." But Langmuir and Urey were chemists, at least in the eyes of their Nobel Prize committees, and both were outspoken supporters. While Astbury, a supporter turned vociferous opponent, was a physicist.

John Archibald Wheeler, a great twentieth-century physicist, collaborated with Bohr and with Einstein. He understood the power of names and had a way with them: "black hole," "quantum foam," "worm hole," and "it by bit" were his. Kip Thorne and Richard Feynman were his graduate students. Throughout his long career at the University of Texas and at Princeton, he taught undergraduates too. "In my freshman physics class," Wheeler told me long ago, "I can always tell which students will go on in physics and which ones will become mathematicians." Their choice had nothing to do with ambition or ability; they were all ambitious and bright. Wheeler's Rohrschach test was the uncertainty principle of quantum mechanics. Not whether the students could state it, but whether they could stand it. Those who didn't balk became physicists. Those who did became mathematicians. He called it right every time.

In the late 1930s, before the results of experiments were in, before the cyclol controversy became *ad feminam*, before World War II changed science forever, Delta's lacy fabric pattern tested for an affinity: an affinity for explaining the most with the least, for drawing straight lines through scattered data, for ellipses over epicycles. An affinity, in short, for simplicity, and for the elegance of mathematics and logic. *Euclid alone has looked on beauty bare.*

Langmuir—inventor, experimenter, theorist all in one—felt that affinity. Before deciding on chemistry, he studied mathematics in Göttingen and was deeply influenced by the geometer Felix Klein. Though he used all the chemical arguments at his disposal to defend the cyclol theory, simplicity was his clincher (italics mine):

> The cyclol hypothesis introduced the simple assumption that the residues function as four-armed units, and its development during the last few years has shown that *this single postulate* leads by straight mathematical deductions to the idea of a characteristic protein fabric.... These cage molecules explain *in one simple scheme* the existence of megamolecules of definite molecular weights, of highly specific reactions, of crystallizing, and of forming monolayers of very great insolubility.[18]

True, he said this to an audience of physicists, but it wasn't just physics-speak: he nominated Delta for a Nobel Prize in Chemistry that year (no, she didn't get it).[19]

Harold Urey studied in Copenhagen after receiving his Ph.D. The busy Bohr assumed he was a physicist and Urey never set him straight. When he wrote to his thesis director, G. N. Lewis in Berkeley, for help in finding a job, Lewis tried to park him in a physics department. "Lewis likely saw Urey as too mathematically oriented for his liking as a chemist," says Patrick Coffey in *Cathedrals of Science.*[20]

Ross Harrison argued that "successful explanation in embryology had to be measured by 'simplicity, precision, and completeness of our descriptions rather than by a specious facility in ascribing causes to particular events.'"[21]

Niels Bohr felt it as he spun uncertainty into gold. In Einstein's words, "That this insecure and contradictory foundation was sufficient to enable a man of Bohr's unique instinct and tact to discover the major laws of the spectral lines and of the electron shells of the atoms together with their significance for chemistry appeared to me like a miracle…This is the highest form of musicality in the sphere of thought."[22] Bohr agonized over the dark heart of physics. We were "reminded time and again by Bohr himself what the real problem is with which he struggles," his friend Paul Ehrenfest wrote in 1923, "the unveiling of the principles of the theory which one day will take the place of the classical theory."[23]

But for Linus Pauling, mathematics was a tool and nothing more. "Mathematics was fine as a tool," says Thomas Hager in *Force of Nature*.[24] He quotes Pauling, "I could never get very interested in it. Mathematicians try to develop completely logical arguments, formulating a few postulates and then deriving the whole of mathematics from these postulates. Mathematicians try to prove something rigorously. And I have never been very interested in rigor."

Young Dorothy Crowfoot, quaking with trepidation at the prospect of graduate study in x-ray crystallography, wrote to her parents, "It is one thing to appreciate the structures that other people have worked out for crystals—and quite another to be able to work them out yourself. The first requires the same faculties I apply to mosaics—the second requires pure mathematics. It is quite dreadful to think about it." "Despite the good progress she was making," Georgina Ferry says in *Dorothy Hodgkin: A Life*, "Dorothy became downhearted by the end of the first term, convinced that her old bugbear, mathematics, was letting her down."[25] She stayed the course, pursued protein structure for 30 years, and won a Nobel Prize in 1964. She didn't need, and didn't use, pure mathematics in her x-ray work after all. But Delta did.

And Bernal? As a small child, he was fascinated by numbers. But mathematics never held him: nothing, and no one, ever would. He took an interest in mathematical problems from time to time—as an undergraduate he won a prize—but the only system that ever captured his heart, or mind, was communism.

15 WHAT *IS* SHE DOING HERE?

"Oh, Dr. Weaver! If I only had a handsome young man to twiddle the test-tubes for me!"[1] No, said Dr. Weaver. No and no again. This was her nth request.

Delta had made her first request a few months into her grant. "It looks rather as if I have come across a key to the problem of protein structure,"[2] she told Tisdale. In a flash of insight, she'd found a radical but astoundingly simple architecture for all the globular proteins molecules. It fit the data and explained their puzzling properties in one fell swoop. The eminent organic chemist Robert Robinson was interested. Which brought her to the point: would the foundation pay for a personal assistant, a chemist, to do experiments for her? He or she would work alongside the other 65 scientists in Robinson's lab.

Tisdale refused. Not because she'd migrated from chromosomes to proteins—that path made sense to him—but because the assistant would only duplicate efforts better done elsewhere. We support protein research in many laboratories in many countries, he told her. While you study their geometry, someone else is doing spectroscopy, another x-ray diffraction, and so on. The truth will emerge collectively, in journals and conferences.

Aside from this general principle, he pointed out, the word "assistant" is a misnomer here. The experiments you want can't be done by a novice. You should cooperate with an existing research group. He, Tisdale, would be happy to talk with Professor Robinson about this. Besides, what will you ask for next—an assistant in genetics?

As per foundation protocol, Tisdale reported the conversation to Weaver. "I think that Dr. Wrinch has the enthusiasm and possibly also some of the impracticality of genius," Weaver replied. "In my opinion you have handled the matter in just the right way." Tisdale was glad to hear it. "I thought it would not only be foolish to think of attaching a protein chemist to her, but I felt that

I would overlook a bet if I did not seize the opportunity to head off similar proposals for a genetics assistant and so forth which as you will observe I did or at least think I did. My reason for expressing doubt is that as you well know my experience in managing the gentler sex is rather limited but so far I have found that the best any man can hope to get in an argument with anyone of them is a half, to introduce a golfing term."[3]

"I earnestly ask you to consider this suggestion sympathetically," Delta implored Weaver, though Tisdale had told her to write only to him. Professor Robinson, she said, had agreed to work on three of the many projects she had proposed for testing her model. He thought the others were futile, but she did not. A good assistant could help her firm them up and flesh them out. "My ultimate objective is to develop the theory to the point at which it can be handed over to the experimentalists. To make out the strongest possible case for these ideas, so that they may seem worthy of the attention of these specialists, is a task of some magnitude." No, Weaver said.

The deeper issue in this standoff wasn't money, or duplication of effort, or handling the gentler sex. It was their very different ideas of Delta's role in the biology revolution. The Rockefeller Foundation was impressed by her mathematical chromosomes; that's why they'd given her the grant. Her protein model, if correct, would turn science upside down. Weaver agreed with her objective, "to develop the theory to the point at which it can be handed over to the experimentalists." But where was that point? And once she'd handed it over, would she stay out of the way?

"There is a great deal of work in front of us before we can bridge the gulf between biology and mathematics," Joseph Woodger had warned Joseph Needham in the Theoretical Biology Club's early days. "We cannot expect them to help us until we have helped ourselves. The sort of questions that we may hope them to help us to answer cannot be asked by them but only by us."[4] Delta was facing this very problem now, but from the other end. She could not expect chemists to help her unless she asked her questions in terms they understood. But just as Woodger and Needham struggled to pose questions to mathematicians, she needed help with concepts and language. She had taught herself to speak chemistry and biology, but she still dreamed in math and physics. An assistant would help triage her ideas and make the best of them palatable. And, tacitly, she reserved the right to judge the experimental results.

Delta had an assistant of sorts. But Eric Neville was a pure mathematician, not an organic chemist. He himself did not deign to apply the noble art of mathematics to any practical problem, but he cheered Delta on for the rest of his life, and she relied on him for sympathy and advice.

He was "tall, loosely built though not athletic; with iron-grey hair...He wore rimless spectacles," Eric Neville's colleague Walter Langford told friends who gathered at Reading University in 1978 to celebrate his donation of his late friend's books to the library. "Behind those glasses, the eyes were often glowing with the light of a puckish humour."[5]

As a student at Trinity College, Cambridge, Neville had been deep in preparation for the 1910 Mathematical Tripos gold crown when he learned that from that year forward the 30 Wranglers would not be ranked by their scores. Loath to lose his chance at immortality, he took the exam a year early. The stakes were high. Over the years, the Tripos had become a contest between Trinity and St. John's next door: going into this final round, these two colleges boasted 55 Senior Wranglers each. "Who will be the last Senior Wrangler at Cambridge?" one headline brayed. Neville came in second; he saved the news clippings all his life.[6] He went on to a Trinity Fellowship; his fellow Fellows included Hardy, Littlewood, and Watson. But then the Great War broke out and changed everything.

Neville worked in a London hospital in the war years. "Although his eyesight would have barred him from any combatant service," Langford continued, "his whole nature revolted at the thought that, in any circumstances, man should kill his fellows, and he declared that as his conviction from the beginning....One can only conjecture whether his Fellowship might have been renewed had he not declared his opinion, but the fact is that it was not and he moved from Cambridge never to return." Neville took his pride with him. At the red brick University of Reading, so near and yet so far from Cambridge, the embittered, taciturn last Second Wrangler cast pearls before swine. Respected in the wider British mathematical community, he served stints as president of the Mathematical Association and chairman of the British Association's Committee on Mathematical Tables. Neville married Alice Farnfield in 1913; their only child, a son, died in infancy. Alice Neville died in 1956 and "those of us who knew him closely have the feeling that he never reconciled himself fully to her passing." But, admitted Langford, E. H. was "by no means an easy person to know or to understand really well."

Like Delta, Neville was close to Russell in the war years and acquired elements of his distinctive style. A reviewer of Neville's forbidding *Prolegomena to Analytical Geometry in Anisotropic Euclidean Space of Three Dimensions* (1923) blamed Russell for Neville's "independence which shows itself in nomenclature and notation, in absence of references, and most of all in the limitations which the author has placed upon himself in the selection of his material."[7] The same objections were raised to Dot's mathematical papers by anonymous referees.

After the war, Delta and Neville served together on the Mathematical Tables Committee. Their thirty-one-year love affair, if that's what it was, began in her *annus horribilus*, 1930, the year of John's implosion. She never loved Neville as he loved her, but he would be the one, perhaps the only one, for whom she didn't wear a mask of gaiety. Her letters to Erice, as she almost always spelled his name, are filled with her fears, frustrations, and worries; his with love and advice. Yeats spoke for him:

How many loved your moments of glad grace,
And loved your beauty with love false or true,
But one man loved the pilgrim soul in you,
And loved the sorrows of your changing face.[8]

Time and again Neville set his own work aside to give her a hand: "The author offers her thanks to E. H. Neville for his advice and criticism"; "Reproduced by kind permission of Prof. E. H. Neville, University of Reading, England, who kindly constructed this model."

"I had tea at the Crowthers who have a charming modern apartment on Russell Square," May Sarton wrote to her father, the Harvard historian of science, on Coronation Day, 1937. "Dorothy Wrinch was there in one of her strange simpering, showing off moods, talking about herself constantly. For some reason it made me feel extremely shy and uncomfortable."[9] "Dorothy Wrinch has finally obtained her divorce," George Sarton told his daughter two years later, "but that has not solved all of her problems, and it will be difficult for her to reconcile her scientific duties with her duties to Pam. The difficulties lie in her own nature."[10]

Delta's friends bought Pam a membership in the British Association for the Advancement of Science, the first ever for a child. Together Delta and Pam spent pleasant weekends with Charles and Dorothea Singer in Cornwall. Dorothea would be one of Pam's legal guardians in the event of Delta's premature death. The Singers were eminent science historians; the Needham-Singer-Sarton connection was bristly but tight. When Delta traveled abroad, which was often, her friends called on Pam at St. Christopher's. Olga Taussky brought her a present. Pam told her to leave it and go away.[11]

The Wrinch-Robinson collaboration began well. Tisdale talked with the professor, as he had said he would. "R tells me frankly that when Dorothy Wrinch first started her conversations with him he was very much prejudiced against her because he could not believe that one with her background could contribute to studies in protein chemistry. He feels quite certain that she could have no place in the direction of laboratory work, but he confesses that to the shame of chemists, Dorothy Wrinch's work has pointed a new direction of further experimentation which offers very fertile fields of investigation."[12]

To Tisdale's surprise, Robinson asked for money for assistants. Was this prompted by Wrinch? Tisdale wondered. Robinson assured him it was not. If the foundation would pay for three more assistants (he had 10 at the time), Robinson would turn some of his present workers toward proteins as well. He planned to start an x-ray group too, headed by Bernal's student Dorothy Crowfoot. "I entered my conversation with R. with a certain amount of skepticism," said Tisdale, "but I left very much enthused with the possibility of a grant of approximately £1500 per year for five years being an extremely profitable venture. R. is without question the most outstanding synthetic chemist in Britain, if not in Europe."

But three months later Weaver sensed tensions. "It is clear that R. is not willing (considers it beneath his dignity probably; and one must agree properly so) to have his department serve as a mere testing bureau for W's ideas, subject to her frequent and perhaps fluctuating enthusiasms. R. is putting several of his own men on this general program of protein synthesis and structure investigation."[13]

Within a year, the relationship was fraught. "I had arranged for Robinson and Mrs R to come to tea to talk seriously about proteins," Delta wrote to Sobotka.[14] "Then Langmuir, plus Mrs and Miss L…decided to visit us at the same time. The result was that R stalked in full of his own importance, only to find that he had to listen for three hours and more to what Langmuir thought he should know."

Imagine the scene: Delta's book-lined sitting room, teacups and sandwiches piled on the table. Two high-brow, high-maintenance, mountain-climbing titans of chemistry face off. Their wives and children and Delta and Pamela look on—for three hours.

Robinson, no stranger to controversy then or later, was also known for his "admirable, but sometimes bewildering propensity to drop an hypothesis the moment he believed it to be incorrect, no matter how beautiful it was."[15] While Delta was known for her propensity—bewildering to most but admirable to others—to retain an hypothesis in the face of counterevidence, partly because of its beauty.

They hit a wall. Did they climb over and shake hands? Not on your life. Robinson didn't need Langmuir to teach him organic chemistry. His experiments had disproved the cyclol model, and that was the end of it. "I hope you won't consider these comments as *lese-majeste*," Delta continued her letter to Sobotka,

> but his [R] childish and idiotic method of arguing really are more than I can stand....He could not see that there is an essential difference between the arrangement of points in a linear sequence and an arrangement of points in a two-dimensional pattern. He would not allow that the difference between the polypeptide [chain] view and mine was roughly the difference between a fishing line and a fishing net. He didn't seem to think that this way of looking at things meant anything at all, and was quite unable to see the geometrical implications of the distinction.

Neville was also present at the tea. The less you have to do with Robinson the better, he advised Delta afterward. Neville had, she went on, "got, in fact, an awfully bad impression, not only of his style of arguing but also of his personality. He thought that one could not trust him in the very least, and that he would take every opportunity to point out that, in spite of Langmuir's techniques etc being introduced into his lab at my wish nothing in favour of cyclols had come out of it, and so [he, Neville] thinks really it is far better not to encourage this project."

If Robinson wouldn't help her, maybe others would. Delta wrote to Weaver from Zürich a week later,[16] "I am here to have talks with Niggli, Rǔzička, and Karrer." Paul Niggli was one of the world's leading crystallographers. Paul Karrer would win the Nobel Prize for Chemistry that year, 1937, and Leopold Rǔzička two years later. (Robinson won it in 1947.) "Rǔzička has been *extremely* helpful. He made many suggestions and has definitely offered to produce certain compounds for me."

Never had Delta listened so selectively; never had she so misjudged and misgauged. News may not have traveled at the speed of light yet, but gossip did. Whatever she said or didn't say, whatever Rǔzička said or didn't say, the story lives on—and on and on—as Robinson told it.

He told it to Tisdale.[17] "R. says he is coming to feel that W. is not playing the game fairly. After he has spent much time, thought, energy and material supplies on her suggestions at her request, with an adverse decision, she approached Rǔzička to do the same or similar experiments. Rǔzička asked why she came to him when she has Robinson in Oxford, to which she replied, 'Oh, Robinson is not interested,' instead of telling the facts of the case."

But, Tisdale added, "this can be said of Robinson, however, he gives the devil his due, or her in this case, he says there is no denying the stimulation she has given to the problem of protein construction, although he does wonder if her changing value has not already passed the apex."

Tisdale wrote his final entry on the saga six months later: "R tried to make some of Wrinch's cyclols, but no success. He and W. have drifted far apart."[18]

In New York before returning to England, her first visit to America drawing to a close, Delta called on Weaver to thank him for making the arrangements and to wish him goodbye. She had visited laboratories and lectured at universities throughout the northeast and talked with many scientists, she told him. Everyone had been very kind. Research projects were blooming or about to. The *New York Times* foresaw benefits for mankind. She would come back as soon as she could.

Weaver listened without enthusiasm. Slow down, he said. You have cast your bread upon the waters—papers in the eagerly read weekly *Nature*, a demonstration in the Royal Society's hallowed halls, lectures. Now it is time to "bend all [your] efforts towards obtaining more explicit and convincing formulation and proof of [your] own theories."[19] Do what you do best; your ideas will find their natural audience.

Usually oblivious to subtexts, Delta caught this one: "this advice has, at least in some part, resulted from criticisms." H. M. Miller, Weaver's colleague, had heard the buzz on a recent trip to England. Wrinch had been going about demanding that everyone drop what they were doing and test her cyclol theory. She was driving them crazy. "She is both alarmed and distressed," noted Weaver, "and there is some imminent danger that the conversation descend to a rather painful emotional level. WW succeeds, he thinks, in calming her down, but in nevertheless convincing her that she must not take so militant an attitude. The question of rapidity of publication also enters the picture, and W. agrees that the time has now come when she can afford to be more conservative."

But how could she, working alone, obtain "more explicit and convincing formulation and proof of her own theories"? If only they would give her an assistant...

I, Dot's unofficial, unpaid assistant, was the last in a very short line. She gave me various tasks, not all of them related to the book we were writing.

She asked me to analyze the hexamethylenetetramine molecule. The tetra and the hex; not the methyl; its geometry, not its chemistry. Our molecule was a sturdy Platonic cage, four hexagons joined three at a corner (Figure 7, left). The hexagons don't lie flat: they zigzag up and down as you circle around. Chemists called them cyclohexane rings.

SKELETON STRUCTURE IS FOUND IN PROTEINS

Key to Body Changes Is Seen in Finding by Dr. Dorothy Wrinch in British Journal.

ROCHESTER, N. Y., Nov. 29 (*P*). —What may prove to be a major development in science, discovery of a skeleton-like structure in the proteins, which compose the principal part of the living human body, is reported in Nature, British Journal of Science.

Dr. Dorothy M. Wrinch of the Mathematical Institute at Oxford finds that protein molecules, which are known as the "building blocks" of living matter, are all alike on one side, their structure on that side resembling a lattice such as is set up in a garden for climbing vines.

Chemists here are interested in the possibility that Dr. Wrinch has discovered a principle of great practical use in chemistry.

"It may prove the key to an understanding of general physiological action of organic compounds and to a great deal of non-physiological chemistry in the case of the high molecular substances now of growing importance," one of them said.

It has been known that there is an almost endless number of different proteins, chemically different, composing the human body.

Dr. Wrinch's discovery that all are the same on one side gives a new idea of the unity of the body. It appears to explain for the first time, in a simple way, the operation of the apparently confusing chemical changes which give rise alike to health, disease and death.

Dr. Wrinch demonstrates that because these proteins are alike on one side a great variety of the other life chemicals can attach themselves equally well, anywhere, to the protein structure. Thus, she points out, nine substances discovered in the last four years related to coal tars, all of which cause cancer when rubbed on the skin, will fit and attach themselves perfectly to the similar "backs" of protein molecules.

These nine cancer-causing substances are artificial and have not been found in the human body. But they are similar in some ways, Dr. Wrinch explains, to two very well known human body chemicals, the sex hormones and the bile acids.

The New York Times
Published: November 30, 1936
Copyright © The New York Times

"Skeleton Structure Is Found in Proteins," *New York Times*, November 30, 1936. Used with permission of the Associated Press, © 2012. All rights reserved.

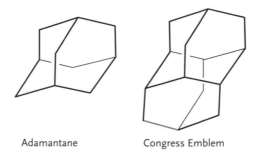

Adamantane   Congress Emblem

Figure 7 Left: The hexamethylenetetramine, or adamantane, cage, is built of cyclohexane rings. Right: The emblem of the XIXth International Congress of the Union of Pure and Applied Chemistry, 1963.

The hexamethylenetetramine cage is also the framework for adamantane; the word comes from *adamantinos,* the Greek word for "of the diamond," meaning "very hard"; it's the ancestor of the English word "adamant." The crystal structure of diamond is built of this cage. Each cage shares its rings with its neighbors. You can't say where one starts or ends. I built models of this intriguing cage, analyzed its symmetries, and drew projections from different directions. Dorothy had a lifetime supply of special graph paper that she'd designed herself. Instead of the usual square meshes, some sheets were covered with rectangles, others with hexagons, all scaled just right for drawing projections of cubes.

One day she handed me the program announcement of the XIXth International Congress of the Union of Pure and Applied Chemistry. The meeting had been held in London in 1963, eight years before, but she had saved the announcement. Its description of the congress emblem had gotten her goat. This diagram, said the circular, represents a "beautifully symmetrical molecule which has not, so far, been described in the literature. If adamantane be regarded as an 'adamantalogue' of cyclohexane, then the Congress Emblem is an adamantalogue of adamantane."

"See what you can make of this," she said. "Give me a copy of your statement and diagrams for keeps." I did and kept my copy too: description, diagrams, models, computations. It was easy to show that "adamantalogue of adamantane" made no linguistic sense, so why had she asked me to do this? I suspect my report was a round of ammunition in her endless war on chemistry jargon.

"LINUS AND DOROTHY"

THE OPERA, WITH TALKBACK

Dorothy Wrinch's epic battle with Linus Pauling is the stuff of opera. There is no other way to tell it. Two brilliant, arrogant, competitive antagonists with a flair for publicity and a touch of the devious! And what a plot!

Act I. 1936. Dorothy, a beautiful, young British mathematician, proposes a model for protein architecture, a simple cage of such Platonic beauty that chemists of all stripes drop their research and rush into proteins.

Act II. 1938. Linus, an ambitious, young American chemist, appalled by her claim that geometry trumps chemistry and by her affinity for the media, publishes an erroneous list of her errors and drums her out of the field.

Act III. 1946. Dorothy, exiled to a small New England college, insists that beauty is truth and truth will out. She defends her model against the mounting evidence. Partial—very partial—vindication arrives too late. She collapses to a footnote in the history of science.

Act IV. 1968. Linus, now twice a Nobel laureate, claims that chemistry trumps medicine. He defends his views on vitamin C, against the mounting evidence, for the rest of his life.

This chapter is a synopsis of my imagined Act II. The librettist will reconstruct the dialogue; here it is largely verbatim from Rockefeller officials' diaries, science journals, and the Wrinch papers. I've compressed a few scenes, paraphrased a few speeches, reconstructed missing fragments, incorporated echoes of memory, even invented a little. But the gist is detailed in the sources. Please hold your questions till the end.

The lights dim as the curtain rises. We hear, but until scene 4 we will not see, the chorus, a hundred or so male scientists, their wives, and girl-friends. They sing just one song (music yet to be written), "The Laughing Song," a song with just one word, "ha ha ha ha…" They sing it softly in all the scenes, louder between them. "The Laughing Song" is a fugue in changing keys.

Scene 1, late January 1938.[1] Linus Pauling's office at Cornell University in Ithaca, New York. He has just completed a lecture series on the nature of the chemical bond. Linus sits alone at the cluttered desk in the visitor's office in the Baker Laboratory. A small table is piled high with papers and Tinkertoy molecules; the blackboard is scribbled with chemical formulas. Through the tall window behind the desk we see the snowy campus, students bundled up against the cold. Linus is 37 years old, his hair already wispy. An industrial chemist has asked him to analyze a small crystal of some unknown stuff. He holds it in his hands, turns it this way and that, mutters, "What is this made of?"

The chorus fades. Linus puts the crystal down and raises his head, but does not rise, to sing his first aria, "Why did I agree to see her?" Irritated, he tells of his recent visit to the Rockefeller Institute in New York City, where Bergmann and Mirsky grilled him about Dorothy Wrinch. He hasn't fol-lowed her work closely, he tells them, but he is "adversely inclined" to it, and to her, because she makes heavy use of geometry and symmetry. Also, she is getting too much publicity. Her ideas are "vague, speculative, and per-haps not very important if true," yet she lures journalists like a Venus flytrap. Look at the ballyhoo in the *New York Times*!

His remarks reached Warren Weaver through the grapevine. Warren gave his pal a call. Wrinch was eager to see him, he told Linus, and he was eager to get them together. She was coming to America again in January. Would Linus grill her and report back to him? You alone "will not be in the slight-est awed by her facility in mathematics and mathematical physics," Warren says, because, frankly, you don't give a damn. Nor will you be bulldozed by her grandstanding: it takes one to know one.

Linus had said yes: he needs Warren as much as Warren needs him. But just as Arthur Stanley Eddington didn't have to go to Africa to choose between Einstein and Newton, Linus knows this interview will be a waste of his time. His mind is made up already, and he still has to pack up his office.

In just a few days, he and Ava Helen will go back to Pasadena and their four children, tended by a neighbor these last few months. The baby must be crawling already!

Dorothy Wrinch raps on the door and strides in, briefcase in one hand, cigarette in the other. *Professor Pauling, I presume?* She is 43 years old, her curly hair dark red, her voice a tad too cheerful, her smile a shade too bright. She takes a seat, *the* seat, the only vacant chair in the office.

Time is short, Linus says, so let's get straight to work. Dr. Weaver wants to know on which points we agree, on which we disagree, and why. I propose to take notes, which you will read and correct.

Delta spots the crystal on his desk. "An octahedron, I see!"

Pauling ignores her and asks his first question. What *exactly* is your theory of protein structure?

Delta sings her theme song, "The Mostest from the Leastest." From one simple postulate, the cyclol bond, all proteins follow! That is what interests her: the rigorous deduction of consequences from postulates. She will abandon her cyclol postulate if it is shown to be wrong. (Linus picks up a pen and writes furiously.) To show her flexibility, her openness to new suggestions, she has just sent a letter to *Nature* on cyclols with hydrogen bonds.

But, says Linus, we chemists don't care about postulates, we care about proteins. Your "elasticity is made possible by the indefiniteness of your conceptions."[2] Your "feelings about science are completely different" from ours. Why don't you just publish your mathematical ideas for what they are, and then ask, dispassionately, whether they match reality?

He bombards her with technical chemistry questions. What exactly holds your cyclol rings together? Doesn't the water trapped in the cages affect their molecular weight? Hasn't x-ray diffraction proved that proteins aren't cyclols at all? (The librettist will convey the drama: not the details but the bombardment. Linus sees Dorothy as a mathematician masquerading as a chemist and aims to unmask her.)

Dorothy defends herself at first but grows unsteady. It's his office, his territory, his ground. As the interview proceeds, her confusion mounts in lockstep with his swagger. Linus completes his notes with a flourish. She reads them, makes a few changes, signs her name, and leaves. He picks up the crystal again.

Scene 2, a few weeks later.[3] Warren Weaver's office at the Rockefeller Foundation on Fifth Avenue in New York. This office is large and flush and plush. Life-sized oil portraits of John D. Rockefeller Sr. and Jr. gaze at each other from opposite walls. Warren's handsome desk and armchair take up a quarter of the space, the polished wood conference table a half. Warren and Tizzy (as he calls Tisdale) sit on one side of the table facing Linus and an unidentified fourth man. (Unidentified, because I have invented him; I need a devil's advocate.) A female secretary sits in a corner quietly taking notes.

The atmosphere is convivial, even jovial. Warren lights a cigar, Tizzy pours a round of sherry. Warren thanks Linus for taking the time to talk with Wrinch and send him their co-edited notes. And now, says Warren, we want your *candid* opinion.

"My impression of her and her work is in general an unfavorable one," says Linus. "I believe, however, that my conclusions are justified. I doubt that her attack on the problem of protein structure will lead directly to any valuable results. It would not be worthwhile to have her working in our laboratories in Pasadena."[4]

Warren and Tizzy exchange nods. Go on, says Tizzy.

"I first attempted to find out if Dr. Wrinch had a definite theory of protein structure differing from the usual polypeptide-chain theory," Linus continues. But she says "she is interested in the rigorous deduction of consequences from postulates rather than in the actual structure of proteins, and would abandon the cyclol postulate if it were shown to be wrong." Very elastic!

But some people, the fourth man points out, criticize her for *not* abandoning her postulate. It seems she's damned if she does and damned if she doesn't. Linus ignores him.

Is there a chance she'll produce something of value? asks Weaver. No, says Linus. "I feel that there is no chance that she will make a significant contribution to the protein problem. It is evident now that geometry and symmetry are of subsidiary importance, the important role being played by interatomic and intermolecular forces." Protein structures cannot be "correctly predicted from a priori considerations."

But, says the fourth man, people tell us her cyclols explain all the facts. "Unfortunately," replies Linus, "the physiologists and chemists who are interested in proteins do not know that her feelings about science are completely different from theirs, and so they are deluded into taking her work seriously."

"Dr. Wrinch is facile in the use of the terminology of chemists and biologists," Linus says, "but her arguments are sometimes unreliable and her information superficial. She attempts to strengthen her arguments by quoting x-ray results without understanding them."

"Moreover," he goes on, "in her papers and in her talks she is, or seems to be, dishonest in her attempts to lead her readers and auditors to believe that proteins have her cyclol structures."[5]

Weaver perks up. Dishonest? In what way?

On a technical point, says Linus. She predicted two different molecular weights for the same cyclol molecule but wouldn't admit it.

Thank you, Linus, says Weaver. She'll want us to renew her grant, and opinion is so divided. "Your comments will be extremely useful to us."

"Almost all of the chemists and physiologists with whom I talked have a dislike for her and her methods of work," Linus adds.

Tizzy pipes up: "I know full well that certain other people do not like her either. On the other hand, viewed more largely, she certainly has been a catalyst, although I am not sure how long a catalyst should continue to exist after reactions have been catalyzed."

"My belief," says Linus, "is that Dr. Wrinch is not contributing greatly to science by her influence on other people."

But, he adds, not realizing, or not admitting, that she has catalyzed *him*, "since talking with her it has seemed to me that I should prepare a paper on my ideas. I have been averse to doing this because of their speculative nature; they are, however, superior to Dr. Wrinch's, in my opinion."

Weaver urges him to write up his own ideas. "One has, of course, to recognize the risks involved in a paper which must necessarily be somewhat speculative in nature; but it would certainly seem to me that the very substantial amount of rigorous and experimental publication which you have to your credit would, without question, bear the burden of a little speculation."

The meeting ends pleasantly. Linus, Tizzy, and the fourth man leave. Warren turns to his secretary. Take dictation! "A confidential report to the Board of Trustees."[6] Facing the door, not his secretary, he sings.[7]

An interesting approach to the whole problem [of proteins] has recently been made by a mathematician, Dr. Dorothy Wrinch of Oxford University.

Her approach is unusual, not only in that it is an effort to apply mathematical ideas to the interpretation of life phenomena, but also in the fact that it bravely tackles the molecule as a whole.

Instead of breaking down the unknown complexity, and then from a study of its parts trying to reason what the original structure must have been, Dr. Wrinch takes an entirely different tack.

In effect, her method is an appraisal of probabilities.

Molecules have certain spatial characteristics which, even though unknown, surely are subject to the mathematical laws of geometry and symmetry.

If we apply these notions of geometry and symmetry to the question of protein form, certain structures immediately are ruled out as illogical; and others soon reveal themselves as untenable on the grounds of the known experimental facts.

In other words, a mathematical theorem can prove that certain combinations or procedures are impossible.

That is progress, for it enables you to throw aside many items on which an experimentalist, working on the trial and error principle, might waste time.

It enables you to list the possibilities, to narrow the list to the probabilities, and thus to arrive at structures that may be regarded as practicable candidates for testing.

Less than three years have passed [since her grant began] and in that time the interest of workers in protein research has showed itself in increased activity in many directions.

It is only fair to say that Dr. Wrinch's hypothesis has met with attack as well as with support, but attack and defense are both forms of activity; and, irrespective of the final outcome of the hypothesis, the total effect of this small investment appears to be all to the good.

The field is one of the most fundamental in biology, it is of very great human significance, and one can say that it is being cultivated today as never before.

Scene 3, the same date as Scene 2.[8] A coffee lounge in Irving Langmuir's laboratory at General Electric in Schenectady, New York. Scientists in white lab coats mill around the coffeepot, browse at the bookshelf. Dorothy, Langmuir, and Langmuir's right-hand-woman Katharine Blodgett sit on cafeteria chairs, drinking tea, and sharing a box of Fig Newtons as they examine a model of Dorothy's cyclol cage.

Katharine works with Langmuir on thin films, though not on proteins. She has just invented invisible, nonreflecting glass. Her father had worked with Langmuir before her; to follow in his footsteps was always her dream. But first she studied with Rutherford in the Cavendish. Her Cambridge Ph.D. was the first in physics for a woman.

Irving Langmuir, Dorothy Wrinch, and Katharine Blodgett examining Dorothy's protein model, November 1936. The Dorothy Wrinch Papers, Smith College Archives

"We had a very good meeting," Dorothy tells them. It was cold and Ithaca is far out of the way, but I am very glad I made the trip. Now the Rockefeller people know exactly where things stand between Pauling and me. I expect to work with him in Pasadena, if the occasion develops.

Watch out, says Langmuir. My friends in California had doubts about his character even before that nasty business at Cal Tech over the chairman's death and the battle for succession. They say Pauling is arrogant and rides roughshod over people who get in his way. He tolerates no dissent. And no dissenters either.

I stood my ground, says Dorothy, except on two points. He said the cyclol cage must be full of water, so its molecular weight must be greater than I'd thought. I recalculated...

But, Langmuir interrupts her, there is *no* water in the cyclol cage! Some of the residues are hydrophobic. That's why the fabric folds up: to keep them away from water!

Dorothy and Katharine stare at him. *That's it!* they shout in unison. *You've found the missing mechanism!*

Langmuir, elated, seizes Dorothy and Katharine by the hands. They jump up from the table and, joined by three scientist bystanders, improvise a hexagonal dance, "The Cyclol."

Indeed, Langmuir has identified a driving force in protein folding, valid for chains and fabrics alike. Beyond this trio, for the next 20 years only Bernal will grasp this. But the "hydrophobic factor" is a cornerstone of twenty-first-century protein chemistry.[9]

The dance ends, the loungers drift away, and our exhausted trio sits down again.

Pauling's other sticky point was x-ray diffraction, says Dorothy. I must look into that.

Yes, you must, says Katharine. It's the only way to prove your cyclols are right.

Scene 4, 1940. The opening reception of a scientific conference at a large American university. Scientists—all male, slide rules dangling from their belts—mill about with their wives, slapping friends on the back, nibbling trendy Ritz crackers with Velveeta cheese, and drinking wine from paper cups. Most hold, in their hands or tucked under their arms, the July 1939 issue of the *Journal of the American Chemical Society.*[10] Everyone is laughing. They, we realize, are the chorus.

The spotlight falls on a knot of half a dozen men bent over in hysterics as they read aloud from the journal. "The Structure of Proteins," by Linus Pauling and Carl Niemann, "was an overnight success. It became biochemistry's *cause célèbre*, adding yet another trophy to Pauling's expanding scientific kingdom."[11] No, this is not the paper Warren asked Linus to write.

"We have reached the conclusions that there exists no evidence whatever in support of this hypothesis," splutters one scientist, gagging on a cracker.

"Listen to this!" another giggles, "We wish to point out that the evidence adduced by Wrinch and Langmuir has very little value."

Pauling really got her number! a graduate student snickers to his beaming girlfriend.

"There can be found in the papers by Wrinch many additional statements which might be construed as arguments in support of the cyclol structure.

None of these seems to us to have enough significance to justify discussion." The chubby reader chortles breathlessly.

Everyone always said her arguments were fishy, another scientist exclaims. I've never bothered to read her papers; now I don't have to! Linus says it all! "The fact that many other parameters were also assigned arbitrary values removes all significance from their argument."

"…the arguments are so lacking in rigor, and the conclusions are so indefinite…"

Only one man demurs. "Its not for nothing he's called the Appalling Linus," he mutters to himself.[12]

The spotlight moves to another knot. A scientist bewails a missed opportunity.[13]

"I was with this group of men several nights in succession at Gibson Island and our one topic of conversation was working out how we would down Wrinch when she gave her paper. We were actually planning out all of the details of the attack on her."

"And then?"

"She just didn't show up."

Now a cluster of women lights up. They are housewives; whatever their dreams at Smith, Vassar, or Bryn Mawr may once have been, none now practice a profession. They sing in cacophony, not harmony.

"Did you hear about her and Eric Neville?"

"They say it's a Canadian biologist now."

"Did you see that ridiculous photo in *Time*? With the headline 'Scientists Bow to Dr. Wrinch'"?

"That poor child! Parking her in boarding school, or with friends, while she gads about the world."

"I have nothing against women scientists, and neither does my husband. He thinks the world of Dorothy Hodgkin."

"Of course. Dorothy Hodgkin is a lady. She works in the lab, and then she goes home."

"Still, it's not good for her children."

"Dorothy Hodgkin has a rich husband, and maids and nannies."

"Dorothy Wrinch is not a lady. She brings her troubles on herself."

Trumpets blare: all turn and face stage right. The door of the hall slowly opens.

Linus Pauling and his wife, Ava Helen, arm in arm, stroll confidently into the room, nodding like a king and queen to courtiers right and left. The chorus cheers, applauds resoundingly.

A tall, self-evidently eminent chemist, Alexander Todd, strides toward the royal pair. "I must say that I derived enormous enjoyment from reading 'The De-bunking of Wrinch!'" he announces with a strong Scottish burr.[14]

The chorus laughs in riotous unison. "It really was high time that somebody put the case against the cyclol theory in definite terms. As far as I can see, the case is unanswerable."

Letters have been pouring in, Pauling replies, all in the same vein, and many requests for reprints. I trust the cyclol controversy is over.

"The Laughing Song" grows louder.

Then Dorothy Wrinch walks through the door, holding 12-year-old Pamela by the hand. They face the Paulings and Todd.

I have already answered all your supposed arguments, point by point, and in print! Delta says. They have more holes than a cyclol fabric. She sings:

> You might read the message in my papers,
> That no molecules with thousands of atoms and well-defined
> structures
> Can be dealt with except by taking in turn all possible structures
> Theoretically constructed, and testing them in turn.

> I should read your work with care in case, by some extraordinary
> chance,
> Some of your brute facts have structural significance,
> But I should not give you the benefit of chatting with me.
> God help biology if you are the type of person chosen to break new
> ground!

Pamela hands a letter to Pauling. As he reads silently, her childish scrawl is projected on a screen.

Bachmair.

Dear Dr Pauling

Your attacks on my mother have been made rather to frequently. If you both think each other is wrong it is best to prove it instead of writing disagreeable things about each other in papers. I think it would be best to have it out and see which one of you is realy right. There are many quarells in the world alas!! Dont please let yours be one it is these things that help to make the world a Kingdom of misery!!

Yours

Pamela P. Wrinch

Pamela Wrinch to Linus Pauling, undated but evidently 1939 or 1940. (The letter was not sent.) The Dorothy Wrinch Papers, Smith College Archives. Copyright: unknown.

The stage lights fade as the chorus resumes the "Laughing Song" and the curtain falls.

Now let's have talkback, or backtalk. Are there any questions?

Why did you put that crystal in Scene 1?

To illustrate a fundamental feature of the cyclol controversy: when a chemist looks at an object he asks what stuff it is made of, while a mathematician looking at the same object asks what abstraction it's an instance of. Linus and Dorothy seem born a chemist and mathematician, respectively, and their childhoods reinforced those inborn traits. Linus's father was, among other things, a pharmacist. Pauling *père* "showed his son the business, bringing him into the sanctum of the drugstore's back room, where Linus played with the Indian skulls stored there as he watched his father making extracts of roots and leaves, or salves by working chemicals into lard, or carefully measuring, mixing, and packaging powdered herbs on a marble slab."[15] At 16, the age when Delta crammed for entrance to Cambridge, Linus entered Oregon Agricultural College (now Oregon State University) to become a chemical engineer. In his first two years, he "learned how to use a forge, hammering horseshoes and hammers and a knife out of red-hot iron; he learned the mining chemist's craft, blowpipe analysis, and fire assay." While in her first two years of college, halfway around the world, Dorothy kept on cramming (now for the Tripos), sat in on Bertrand Russell's logic lectures, and played the piano in her spare time. She pondered the subtleties of complicated integrals; he imagined atoms of iron hardening into steel.

Small wonder they talked past one another. "We know more than we can tell," explains her friend Michael Polanyi. Knowing is more than rationality, it is also tacit, skills and attitudes we have picked up along the way, "problems and hunches, physiognomies and skills, the use of tools, problems, and denotative language...all the way to the primitive knowledge of external objects perceived by our senses."[16]

But though Dorothy and Linus had deep and real differences, we shouldn't make too much of them. Two years after that interview, he made a remark, later oft-quoted, which sounds very much like Dorothy. "I found that Landsteiner and I had a much different approach to science," he told someone. Karl Landsteiner, a Nobel Prize winner (Medicine or Physiology, 1930), had distinguished and classified the main blood groups. He and Linus were working on antibodies together. "Landsteiner would ask, what do these experimental observations force us to believe about the nature of the world? and I would ask, What is the most simple, general, and intellectually satisfying picture of the world that encompasses these observations and is not incompatible with them?"[17] Dorothy and Linus were more alike than either of them ever admitted.

Why did Weaver make a strong case for Dorothy to his trustees, right on the heels of Pauling's negative report?

I can think of two reasons; there may be more. Like every executive answering to a board, Weaver walked a tightrope. He wanted his programs to succeed, in this case nothing less than the transformation of biological research. For that he had to court, manage, and sometimes deflect, ambitious and cantankerous scientists. At the same time, he had to convince the Rockefeller trustees that their money was being spent well.

But also, Weaver was a mathematician himself. He understood, if Pauling did not, what Delta was trying to do. In his report to the trustees, he approved Delta's *modus arguendi* if not her *modus operandi*: "taking in turn all possible structures theoretically constructed, and testing them in turn." Pauling's remark, that protein structures cannot be "correctly predicted from a priori considerations," did not, it seems, persuade him.

Now, 70-plus years later, mathematical biology is the fastest-growing field in mathematics. Thousands of protein structures are known; you can download the minutest details from an international databank. Yet protein folding, the problem Dorothy hoped cyclols would explain, remains unsolved. "Nature has an algorithm which specifies the three-dimensional structure of a protein from its amino acid sequence alone," says Arthur Lesk, a polymath of molecular biology.[18] "We ought to be able to discover it." The ultimate goal is the "pure *ab initio* approach," prediction from general principles of physics, chemistry, and biology.

Nobel goal? Or hopeless quest, as Pauling thought? Time will, presumably, tell. Meanwhile, says Lesk, suppose you think you've found a method for protein structure prediction. "Whom must you satisfy? The following list is sorted, roughly, in decreasing order of difficulty.

1. Crystallographers
2. NMR spectroscopists
3. Granting agencies
4. Referees of papers
5. Colleagues
6. Your mother"

How do you know what the women were saying? Wouldn't they have stuck up for Delta?

Not likely. It is true I wasn't there, but I've been somewhere very much like it. Dorothy's account, in a letter to Neville, of men plotting to down her when she gave her talk, and Bertrand Russell's remark to Colette that Madame Nicod was "very jealous of Miss Wrinch, because Nicod and Miss W. talk work together," brought back memories of laughter in my father's laboratory and at our family dinner table. Dorothy was not the butt of those jokes. The victim/provocateur's name was Marie.

Marie Nyswander is remembered today as a pioneer in methadone treatment for addiction. She was a psychiatrist, like my father. Marie worked with him for one year, in the mid-1940s, a year nobody ever forgot. Medium height, with dark brown hair worn loose, she was smart and she was lively. She'd studied at Sarah Lawrence, she flew airplanes.

Marie, my mother made sure I understood, brought all her troubles on herself. For one thing, she had a strange sympathy for the imprisoned drug addicts, those con men, jazz men, thieves, and pimps. Impatient with talk therapy and occupational therapy, the treatments then in use, she argued for a chemical cure or, failing that, a less harmful replacement.[19] No, my father said, that would substitute one drug for another.

Marie broke all their rules. At Christmas she gave the inmates the gift they wanted most: morphine shots. "The boys are lonely," she said, "and this will cheer them up." That story, you can be sure, made the rounds for years![20] Yet my father welcomed women in science. Why, Anna worked in his lab. She was German, a chemist, tall and quiet, her dark hair swept up in a bun. My mother thought kindly of Anna, which gave Dad permission to. At one dinner in our home, my sister Norma asked Anna, in front of everybody, "Do you love my daddy?" "Doesn't everyone?" Anna replied graciously, and everyone smiled. Until I Googled "Anna Eisenmann," I hadn't known she'd published papers like the men.

Did Marie love our daddy? Norma didn't ask, which is just as well. My mother, and all the wives, feared Marie was a magnet for men. Marie jokes were vaccines. Their cocktail parties—there were a lot of cocktail parties—bubbled with Marie stories.

The women were jealous of Marie in another, perhaps deeper, way. Twenty years after the ratification of the Nineteenth Amendment, 20 years before *The Feminine Mystique*, educated, middle-class American women spent their long afternoons, from lunch till the cocktail party, playing bridge. They were torn between worlds, the domestic and the wider, but didn't dare to question their predicament or even speak its name. They resented Marie, a free spirit who did as she pleased and lived to tell the tale. So they told tales about her.

Norma didn't forget the laughter either. She became a sociologist and ran workshops on gender stereotyping for judges in American and Canadian courts.

The court of scientific opinion woke up at last in 1975, when Anne Sayre, a writer and editor and crystallographer's wife of a new generation, later a lawyer and judge, wrote *Rosalind Franklin and DNA*, debunking James Watson's caricature of Franklin in *The Double Helix*. Dorothy Wrinch died the next year.

Did Pauling and Niemann end the cyclol controversy?

On the contrary. By using a sledgehammer instead of a flyswatter, they ensured it would be talked about for a century. Two centuries, if this opera ever reaches the stage.

Tone aside, their arguments against Delta's cyclols were dubious. They were right that she was wrong, but they gave the wrong reasons. King James I had been right about smoking, a "custome lothsome to the eye, hatefull to the Nose, harmefull to the braine, dangerous to the Lungs, and in the blacke stinking fume thereof, neerest resembling the horrible Stigian smoke of the pit that is bottomlesse." In his eloquent antismoking tract, *A Counter-Blaste to Tobacco*, the King laid out the arguments, pro and con. Smoking's proponents held that the practice is good for you because smoke is hot and dry, the brains and lungs are cold and wet, so smoking balances your humors. The King laughed them right out of court. *Nonsense!* "The humors of the body should balance *as a whole*, not in each and every part! By warming and drying your lungs, smoking *destroys* your precarious balance!"

Pauling and Niemann raised three objections to the cyclol theory.

From thermochemical calculations (which they gave in some detail), they concluded "that the cyclol structure is so unstable relative to the polypeptide structure that it cannot be of significance for proteins." But, Langmuir complained to Weaver at the time, "in applying the same approximation to another case, I can 'prove' in exactly the same way the non-existence of a molecule whose structure is perfectly definitely known to exist."[21] Even granting approximation wide latitude, their calculations were off by a third, Delta showed them. (They accepted that, but said it didn't matter.) In fact, cyclols do exist: they have been found in hundreds of compounds (though not in proteins, and only in rings, not as fabrics or cages). Confronted with this in late life, Pauling recalled, correctly, that he hadn't said cyclols were *impossible*, only that they were extremely unlikely. But surely he had known, back in 1938, that his readers and auditors would miss that subtle distinction. As I said, he and Delta were very much alike.

Pauling and Niemann's second argument concerned molecular weights: *Wrinch was using the Bergmann-Niemann formula incorrectly!* "We conclude that the cyclol hypothesis does not provide an explanation of the occurrence of amino acid residues in numbers equal to products of powers of 2 and 3." But if Dorothy was wrong, Niemann was wronger: those powers of 2 and 3 were less chemistry than numerology. By stressing her errors, he drew attention to his own and hastened his theory's demise.[22] In using this argument, Pauling was, or seemed to be, dishonest. He didn't believe in numerology any more than he believed in symmetry. "I think the Svedberg rule [on molecular weight classes] is an accident," he had told Weaver, "my idea being that certain protein structures, once started in life, are passed on from generation to generation and species to species."

Third, they said, Dorothy's diffraction calculations were wrong. They hadn't redone them, but Hodgkin and Bernal and others said she was wrong, and that was enough for them. They ignored her important contribution to x-ray diffraction theory, if indeed they were aware of it.

And so we arrive at the heart of the matter. As the historian of science David Berol has pointed out, the cyclol model was a red herring. The real battleground was x-ray diffraction: x-ray diffraction and insulin.

V THE ROSETTA STONE OF THE SOLID STATE

Snow crystals and all the rest besides have much to teach us about the very nature of form.

—D'ARCY THOMPSON, *On Growth and Form*

17 CRYSTALS

I spent my first sabbatical leave at the University of Groningen in the Netherlands. We lived in the mostly rural village of Paterswolde, nine kilometers south of the city, in a red brick farmhouse with tall mullioned windows and a thickly thatched roof. Our farmer neighbors wore wooden shoes in their muddy pastures, and the village children wore them to school. Wooden shoes were warm, dry, and custom-made; the klompen-maaker consulted the orthopedist for problem feet. It is not true that I made my girls wear them to school when we returned to Massachusetts. I only urged them to bring them in for show-and-tell. Don't believe everything you read on the internet.

I went to Groningen to learn How Crystals Get Their Facets. Three days a week, I biked against the wind—always against the wind—from Paterswolde along the shimmering Paterswoldsemeer to Professor Wiepko Perdok's office at the university. Music churning from barrel-organs on Gronigen's streets cheered me on. His office was in an old building in the old part of town, though the science labs had relocated to new facilities in the outskirts.

"Avoid clichés like the plague," warned William Safire, but it *is* a small world. Perdok had learned crystallography from none other than the author of *Lectures on the Principle of Symmetry*, the book that inspired Dorothy Wrinch on March 31, 1923. And Perdok's Leiden colleague, Piet Hartman, had studied with another great Groningen crystallographer, Pieter Terpstra, author of *A Thousand and One Questions on Crystallographic Problems.*[1] Perdok gave me that book and put me through some of the 1,001 questions. He taught me to measure the angles between facets of a crystal with an optical goniometer, a device that bounced light off them. Hartman and Perdok taught me their periodic bond chain theory of crystal morphology and together we predicted the "ideal" shape of apatite crystals, the stuff of bones and teeth, from the bonding of their atoms.[2]

One day Perdok and I drove to Leiden to consult with Hartman, who had just returned from a trip to France. The customs officers had given him a very hard time. What are these strange contraptions, these wooden balls and sticks? Crystal structure models, Hartman explained. I've been invited to France to lecture on crystals. The officers were unpersuaded. Back and forth for an hour; then Hartman guessed the password: "La crystallographie, c'est une science française."

Crystallography *was* a French science for a very long time.

"It seems we cannot doubt that some stones have internal organization; they draw liquid nourishment from the earth; this juice must be carried from their surface, which should be regarded as a kind of skin, to all the other parts," wrote Joseph de Tournefort in 1702. When natural philosophers began to pay attention to crystals, they leaped for analogies. Vegetative theories held sway. Where my friend Mr. Gare, the Northampton jeweler, saw defects, Tournefort saw veins: "there is no more difficulty in imagining this than in understanding how the sap passes from the roots of our largest oaks and pines to the tips of their highest branches."[3] Tournefort, a botanist, is remembered not for his theory on stone growth, but for defining "genus" for plants. The great classifier Linnaeus, who made such use of genus, was born in 1707, the year before Tournefort died. Linneaus's 1768 treatise on stones is best forgotten too: he classified them as he did plants, by their sexual characteristics. Earths were mothers, salts fathers.

But, as Dot insisted to the Aristotelians, myths, analogies, and even brute facts are not yet science. Crystallography became a science in 1773, when a French mineralogist, Jean Baptiste Louis Rome de l'Isle, observed that the different forms a crystal species can take are variations on a theme. All pyrite crystals are cubes, or cubes with their edges trimmed a little or a lot. All diamond crystals are octahedra, or octahedra with their corners snipped. Rome de l'Isle's student Arnaud Carangeot—I name names to emphasize their Frenchness—invented the contact goniometer (a mechanical forerunner of the one I learned to use) to make accurate crystal models from clay. A simple device—just two rods joined at the center of a graded circle or arc—but with it one could *measure*.

Early in the nineteenth century, legend tells us, another French mineralogist, the Abbé Haüy, dropped a friend's magnificent calcite specimen. The treasure tumbled onto the stone floor and shattered in myriad pieces. Haüy, genius at the ready even under these mortifying circumstances, examined the shards. Unlike glass shards, they all looked alike. They weren't identical: some were long, some short, some thick, some thin. But all were brick-shaped

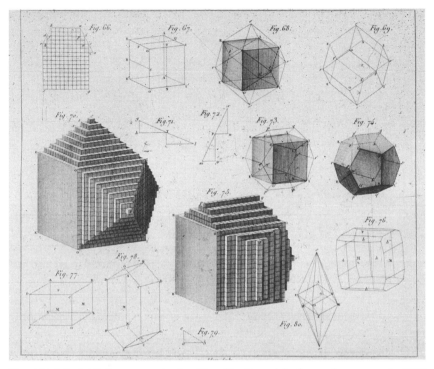

The Abbé R. J. Haüy proposed a building-block model of crystal structure in 1801. This plate, from his influential *Traite de Mineralogie*, shows that different crystal forms can be built with the same bricks.

and the bricks had the same angles. Maybe, Haüy surmised, if we shattered these shards, the smaller ones would look just like them, and if we shattered those…At the end of this infinite sequence of smaller and smaller shards we will come, said Haüy, to the crystal's ultimate building block, its *molécule integrant*.

Haüy used this insight to explain de l'Isle's observation. By stacking the right *molécules* judiciously, he could build each crystal's form, the theme and all its variations. He wrote a treatise showing how to do it.[4]

Haüy's hypothetical *molécules integrants* raised 1,001 questions, give or take a few zeros. The angles of his building-block crystals didn't quite match the angles of the real ones. And what, really, were *molécules integrants* supposed to be? Dalton's atoms? Centers of force? Haüy's student Gabriel Delafosse came to his rescue. It doesn't matter what they are, he said, it's *where* they are that counts. Forget the blocks, use points to mark their centers. Points equally spaced in rows, rows equally spaced in planes, planes equidistant in space: that's the essence of Haüy's idea. A crystal is a page in

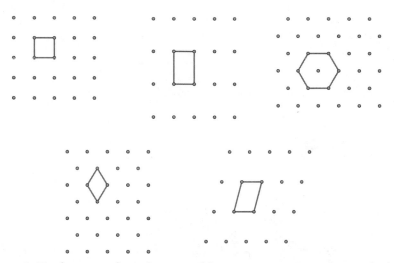

Figure 8 The five types of two-dimensional lattices: square, rectangular, regular hexagonal, diamond parallelogram, and ordinary parallelogram.

a three-dimensional *Grammar of Ornament*, a finite portion of an infinite lattice.

Delafosse was daring. Points? A crystal isn't made of points. How can a point lattice capture a stone's physical, chemical, and geological complexity? (Crystallographers raised similar objections to Delta's work a century later.)

Crystal-minded mathematicians and mathematically-minded crystallographers pushed on. In 1850, in a heroic feat of visual gymnastics, Auguste Bravais proved that there are 14 distinct types of lattices in three dimensions. This gave impetus to the new mathematical subject called group theory. And so the ornate, motif-rich *Grammar of Ornament* became a mathematical text, a catalogue of abstract symmetries.

But "a first rate theory predicts, a second rate theory forbids, and a third rate theory explains after the fact."[5] As a theory of crystal structure, lattices were second rate.

They did forbid. Just as three-mirror kaleidoscopes produce patterns of hexagons, squares, and triangles and no others, the symmetry of lattices is limited. (See figure 8.)

No regular pentagons, or heptagons or octagons or nonagons or decagons…that's why, generations of students were taught, the pentagonal faces of pyrite crystals are always irregular. The lattice theory of crystal structure banished two of Plato's regular solids—the icosahedron and the dodecahedron—from the crystal kingdom. This became known as the "crystallographic restriction." (See figure 9.)

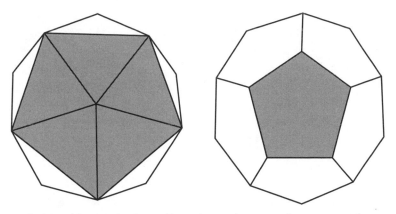

Figure 9 According to the lattice hypothesis, the external symmetry of a crystal must be consistent with a lattice. Because the regular icosahedron and dodecahedron have fivefold symmetry, which lattices cannot, these forms were thought to be impossible for crystals.

Second-rate theories wouldn't do for turn-of-the-twentieth-century physics. Physicists gave the lattice low marks—it was a useful fiction at best—and most of them forgot about it.[6] They were preoccupied with a more urgent question: what are x-rays? In the few years since Roentgen had discovered them, x-rays had revolutionized medicine and physics (and awakened intense interest in the occult). But their nature remained as obscure as teleportation. Are x-rays waves, like light? Or are they streams of particles, like cathode rays? Experiments raised questions instead of settling them. Streams of particles would be deflected by electric and magnetic fields, but x-rays weren't. Waves of any kind can be diffracted, yet, it seemed, x-rays couldn't.

In 1912, Max Laue, a young physicist in Munich, thought maybe they could be.[7] He knew that the dimensions of a diffraction grating must match the wavelength. That's why windowpanes don't diffract visible light but a silk handkerchief does. If x-rays were waves, the wavelength would be so small that no man-made grating would be fine enough to diffract them. But Laue had heard about the old, more or less discarded lattice hypothesis. *If* crystals were lattices and *if* x-rays were waves, a crystal would be a diffraction grating of just the right dimensions. X-rays passing through a crystal would bounce off row after row of atoms, build up like a tsunami, and show up as sharp bright spots on a photographic plate.

Laue asked two students to try the experiment. In their first attempt, the students used a crystal of low symmetry and the photograph was cloudy. Then they tried zinc blende (ZnS), a cubic crystal, and the spots were sharp and clear.

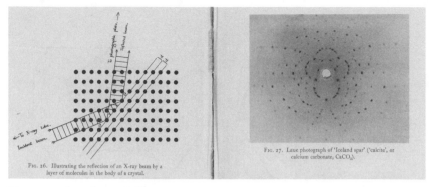

Fig. 26. Illustrating the reflection of an X-ray beam by a layer of molecules in the body of a crystal.

Fig. 27. Laue photograph of 'Iceland spar' ('calcite', or calcium carbonate, CaCO₃).

The principles of x-ray diffraction.
From W. T. Astbury, *Fundamentals of Fibre Structure* (Oxford University Press, 1933), 64, 65.

So x-rays *are* wavelike! And, evidently, crystals do have internal lattices! And, since the waves are diffracted by *something*, atoms must be real (yes, there were still doubters). The experiment was a perfect trifecta. If ever a single experiment "changed the world," this one did. From the double helix to the pentium chip and all that has followed from them, our world has been molded by a fortuitous hunch. Diffraction photographs were the Rosetta stone of the solid state, awaiting their Champollion.

At the turn of the century, William Henry Bragg, a nearly middle-aged British professor of physics at the University of Adelaide, had yet to make a name for himself. Like Bernal and Delta, "he was an adept at picking up a subject, almost casually, making an important contribution, then dropping it again."[8] Then radioactivity caught his attention, he caught the attention of Ernest Rutherford, and the Braggs returned to England. As Cavendish Professor at the University of Leeds, WHB (as he would be called) championed x-rays as particles.

WHB's 19-year-old son, William Lawrence (soon called WLB) went straight to Cambridge and settled on physics. Success came quickly. Laue's team had x-rayed zinc blende in May 1912; in October, WLB puzzled over their paper as he strolled along the Cam. Laue used wave theory to interpret his results; but WLB thought that was too complicated. In a flash of insight that would guide his research for the next half century, he saw True Analogy between x-rays and visible light. Think of the sheets of atoms in the lattice as planes, he reasoned, and think of diffraction as reflection. Everything is illuminated! He rediscovered a simple formula, known now as Bragg's law, that predicts the arrangement of spots in the photograph from the wavelength, the distance between the lattice planes, and the angle of the incident x-ray beam.

But something wasn't right. Zinc blende crystals are cubic, so they should have a cubic lattice. WLB calculated where the spots should appear on the photograph if this were so, but that wasn't where they were. Then he remembered: not one but three of the 14 Bravais lattices have cubic symmetry. One of the three is an infinite stack of cubes, with lattice points at all the corners. Another has points at the centers of the cubes as well. In the third cubic lattice, the extra points are at the centers of the cube faces. WLB calculated where the the spots should be if ZnS was a lattice of the third type. This time, they matched the X-ray diffraction photographs exactly.

WLB's breakthrough was as momentous as Laue's. Crystal structures could be solved in exact detail, each atom assigned its rightful place! He set right to work. First he x-rayed a crystal of ordinary table salt and showed it to be a three-dimensional checkerboard of alternating sodium and chlorine atoms.

Today the checkerboard model sits in every chemistry classroom; it's hard to believe it once shocked chemists. But shocked they were: where are the *molecules*? Where, in this picture, are the sodium and chlorine *pairs*? They don't crystallize in pairs, said WLB. Crystallographers gulped and learned their lesson. Don't think molecules, think lattices. Just as a swatch of wallpaper or fabric tells you all you need to know about the pattern, the arrangement of atoms in a single lattice parallelpiped—a so-called unit cell—is all you need to know about a crystal. Think swatches; think unit cells.

WLB solved the zinc blende structure next, and then father and son solved diamond together. The Nobel Foundation awarded its physics prizes to Max von Laue in 1914 and in 1915 to the Braggs jointly, "for their services in the analysis of crystal structure by means of X-rays." Soon the lattice became the very definition of "crystal." For the next 60 years, no one doubted that "lattice" and "crystal" were—mathematically speaking—one and the same.

In 1923, the now extremely eminent Sir WHB, F.R.S., was appointed director of the Royal Institution in London. As was the custom, he moved with his family into the top-floor flat. WLB, also F.R.S. by then, didn't come with them; he stayed in Manchester, where he was a professor. Under the Braggs' leadership, the R.I.'s Davy Faraday Laboratory and the University of Manchester became world centers of x-ray crystallography.

Bernal and Astbury were two of 12 young scientists WHB brought with him to the Royal Institution. Unusually for the time, the 12 included three women; one of them, Kathleen Yardley Lonsdale, would become the first female F.R.S. in the Royal Society's 300-year history. Sir William Henry Bragg is why there are so many women crystallographers, a later hire told me. Not because we like pretty patterns! Bragg gave us jobs when no one else would.

Astbury and Yardley took on the huge task of preparing tables for crystallographers: tables of lattices and also their intricate mathematical extensions called crystallographic groups. These so-called space groups had been enumerated years before by mineralogists and mathematicians, but no use was found for them so they were put on the shelf. Astbury and Yardley dusted them off, tabulated all 230, devised symbols and notation, and gave the patterns the labels by which they are still known around the world. This standardization was a great help to, and helped to unify, the burgeoning international community of x-ray crystallographers. But it also hog-tied the subject and made it inflexible (until recently; see the epilogue).

Within a decade, x-ray crystallography expanded beyond minerals. "Early in 1926, Sir William gave an afternoon course of R. I. lectures on 'The Imperfect Crystallization of Common Things,'" Dame Kathleen recalled. "He asked W. T. Astbury to assist him in the preparation of this lecture by taking x-ray photographs of natural fibres, such as were being taken at the Kaiser-Wilhelm Institute für Faserstoffchemie. This Astbury did with such thoroughness that he became interested in the field and when an opening occurred in Leeds for an x-ray physicist to study textile fibres, W. H. Bragg persuaded Astbury to go. He needed a good deal of persuasion; none of the workers in the D. F. laboratory ever wanted to go."[9] Astbury broke the sad news to Bernal, who had already moved on to a lectureship in Cambridge: "It seems possible that I have abandoned crystallography."[10] In fact, his career had just begun.

Meanwhile, in Cambridge, Bernal turned x-rays on native protein crystals. He wasn't the first to try it, but no one had yet succeeded because the protein crystals collapsed.

Bernal pondered the problem. Then, suddenly, he understood. The proteins collapsed because they dried out! Suppose he didn't let them? "In the spring of 1934 I was fortunate to obtain crystals of crystallized pepsin…and foreseeing the above difficulties examined a single crystal bathed in its mother liquor and the main outlines of the crystal structure were made out. The cell is relatively enormous."[11] Bernal knew the importance of what he had found. X-rays would reveal the structure of proteins, the secret of life. He walked all night through the streets of Cambridge.

Bernal wrote the words I just quoted for Joseph Needham, for their proposal to the Rockefeller Foundation for an Institute of Morphology. Ironically, his discovery would, within 20 years, do to the problem of biological form what WLB's had done to crystal form: relegate it to the background.

18 X-RAYS AND INSULIN

"Dr. Watson, Mr. Sherlock Holmes," said Stamford, introducing us.
"How are you?" he said cordially, gripping my hand with a strength
for which I should hardly have given him credit. "You have been in
Afghanistan, I perceive."
"However did you know that?" I asked in astonishment.
"Never mind," said he, chuckling to himself. "The question now is about
hemoglobin."

—ARTHUR CONAN DOYLE, *"A Study in Scarlet"*

Exactly, my dear Watson. Hemoglobin, myoglobin, and insulin too.

In Holland I learned enough Dutch to glean the news from the
radio. I ate stamppot, the national veggie-sausage stew, and I liked
it. From colleagues I heard about the intriguing graphic art of their
friend M. C. Escher and fell under the spell of his interlocking birds,
fish, fields, and lakes, the soothing Dutch landscape I crossed every
day on my bike. Dutch scientists used Escher's designs to test their
students, just as I had taught myself symmetry from *The Grammar
of Ornament.*[1] In the old university building haunted by the ghosts
of Professors Jaeger and Terpstra, I measured crystal angles and computed the shapes that crystals should, theoretically, have. And in the
new laboratory outside the city, I tried my hand at x-ray diffraction.

To diffract x-rays with a crystal, you have to have one. The
students in my symmetry course at Smith had grown crystals of
alum and a few others, using recipes we found in a paperback
gem, *Crystals and Crystal Growing.*[2] These crystals were easy to
grow: dump the chemicals in a jar of water, heat the water until
they dissolve, and let the brew cool. Tiny crystals fall to the bottom. To grow a crystal large enough for our simple experiments,
we'd pluck one from the bottom of the jar to serve as a seed, tie a
thread around it, reheat the solution to redissolve the precipitate,

and dangle the seed from a pencil laid across the jar. Now as the solution cooled, the chemicals migrated to the seed instead of falling down. If the evaporation was slow and the lab peaceful, we'd find a large, beautiful crystal hanging from the string in the morning. But easy come, easy go. One student brought her visiting parents to see her magnificent crystal, but the lab had overheated in the midday sun and the crystal dissolved. All that was left was the string.

For x-rays, I needed a crystal that wouldn't melt. Professor Aafje Vos, the lab director, gave me one (I forget what kind) and told me what to do with it: glue it to a rod in the x-ray machine, insert film in the slot provided, close the lid, and flip the switch. For the next 24 hours, the x-rays streamed along the slowly turning rod and the crystal scattered them in all directions. The next day I peeked a few minutes too soon. I survived the electric shock, but the experiment failed. My photographs defied interpretation: I hadn't aligned the crystal quite right. Professor Vos and I agreed I would not try again.

Had my photograph shown bright spots, I would have pored over them, measuring their brightness by eye. Then I would have calculated which hypothetical arrangements of atoms scattered x-rays that way. That's how WLB had tackled zinc blende, first calculating the diffraction patterns of a simple cubic lattice and, when that didn't match his photographic data, trying another lattice instead. Trial and error: that's the best way to do it. By 1974, the year I was in Holland, unanimity on this point was crumbling, but most crystallographers still tried, erred, and tried again.

Solving crystal structures by x-ray diffraction was a two-step process. First, collecting the intensity data. For that you needed an x-ray machine, a relatively defect-free crystal, and film to record the diffracted x-rays. Second, solving the structure: working backward from the diffraction photograph to the arrangement of atoms in the crystal. For that you needed "intuitive powers which border on wizardry—or witchcraft."[3] Although there is a straightforward mathematical relation between the crystal structure and the diffracted waves, you can't observe or measure the waves. The positions and relative brightness of the spots in the photograph do say something, but only what you know about an ocean from the way it laps the shore.

"What most experimenters take for granted before they begin their experiments is infinitely more interesting than any results to which their experiments lead," Norbert Wiener is said to have said. His remark goes straight to the battleground. Most x-ray crystallographers took it for granted, before they began their experiments, that crucial information would go missing. The waves themselves, their ups and downs, would be lost and gone forever. The wizardry of x-ray crystallography consisted of strategies for getting around this "phase problem."

One strategy was akin to using isotope tracers in diagnostic medicine. Crystallographers added "heavy atoms" to the crystal's "mother liquor" to replace certain others in the crystal structure. Because these heavy atoms diffracted strongly, the spots they produced could be identified and would give a first approximation to the structure. With that framework in place the positions of the other atoms could be deduced by trial and erorr. This strategy, called the "heavy atom method," or "isomorphous replacement," reduced the work of the trial and error but didn't challenge it. Two other strategies did challenge it. One of the challengers was Delta.

"Most crystallographers in those days thought the protein structure problem was hopeless," wrote Francis Crick in *What Mad Pursuit*. "The people working on it were optimists though. Because anyone who had doubts would have already left the field." Dorothy Crowfoot Hodgkin never left it.

Dorothy Crowfoot was born in Cairo in 1910 to British archaeologist parents who were usually abroad. The family base in England was the village of Geldeston, whose railway station Delta's grandfather had run. For some reason—class? chronology? chance?—the Crowfoot-Wrinch circles did not overlap there; the Dorothys met in Oxford. Margery Fry, principal of Somerville College from 1926 to 1930, was their mutual mother-confessor. Their lifelong devotion to her was a tie that bound, however frayed.

Dorothy Crowfoot found her calling early; at nine she built a chemistry lab in the attic of her home. After undergraduate studies at Somerville, she joined Bernal's laboratory in Cambridge (and also his stable of lovers). When she returned to Somerville in 1934, Delta welcomed her with a bouquet of roses and helped her through her first, difficult months. "What I should have done this term without her I don't know," Crowfoot told Bernal in December. "I got rather spoilt in Cambridge."[4]

Research was Crowfoot's passion and her strength. Though she tutored Oxford's women chemistry students—Margaret Roberts, later Thatcher, among them—and would be revered as a mentor, she didn't enjoy teaching and, by her own account, wasn't particularly good at it. She set up an x-ray lab in Oxford, with Robert Robinson's help, and published her first paper on the x-ray diffraction of insulin crystals in 1935.

The mysterious substance had been isolated in 1890, but its miraculous powers went unheralded until 1922. It was crystallized that same year. Dorothy Crowfoot was the first crystallographer to try to solve its structure. Insulin was a good choice for a determined, intelligent novice: its molecular weight was relatively low, as proteins went. She succeeded after 35 arduous years.

In her first paper on insulin, Crowfoot presented the facts she'd found, but she drew no conclusions about the molecule's structure. Three years later, Hans Jensen, a Johns Hopkins endocrinologist, lamented that no one had. "The chemical study of insulin has resulted in the accumulation of a large amount of data which at first glance seems to offer little exact information as to the structure of the hormone." In *Insulin, Its Chemistry and Physiology,*[5] he summarized the known facts. But Jensen overlooked, or chose not to discuss, one paper that just made his January 1, 1938, cutoff: "On the Structure of Insulin" by Dorothy Wrinch.[6]

To Delta and Langmuir, Crowfoot's data screamed cyclols for purely mathematical reasons:

- Crowfoot had shown that each molecule in the insulin crystal was surrounded by eight others, and cyclol polyhedra have eight facets.
- The cyclol C_2 fits nicely into the lattice Crowfoot had determined for insulin.
- Insulin has 288 amino acid residues and the C_2 cyclol accommodates precisely that number.
- Delta had shown, mathematically, that a cyclol fabric can fold into a cyclol polyhedron.

Q.E.D.

Delta published these observations in the *Transactions of the Faraday Society* and graciously offered her thanks to Bernal, Crowfoot, Langmuir, Bragg, Sobotka, and others.

Bernal and Crowfoot were not pleased, they were appalled. Delta's interpretation was wholly and hopelessly wrong, they said. Crowfoot promptly published her most recent data on insulin's density, morphology, x-ray diffraction photographs, and molecular weight. "The patterns calculated do not appear to have any direct relation either to the cyclol or to any of the chain structures put forward for the globular proteins," she wrote. Besides, "from the cyclol structure...one would expect many more peaks in the Patterson synthesis than do actually occur."[7]

Which brings us to Norbert Wiener again.

In 1933, Arthur Lindo Patterson, a young British crystallographer trained by Bragg, was working in the x-ray lab at MIT. "I had many opportunities to talk with Wiener," he reminisced. Norbert Wiener was the world expert

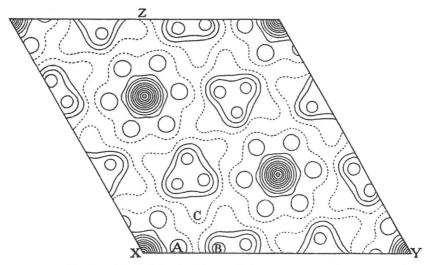

FIG. 1. $P(xy)$ for insulin derived from Patterson-Fourier
analysis. Contours at 5 units apart.

Part of Dorothy Crowfoot's Patterson map for insulin. From Dorothy Crowfoot,
"The Crystal Structure of Insulin. I. The Investigation of Air-Dried Insulin Crystals,"
Proceedings of the Royal Society A, 164, no. 919 (Feb. 1938), 580–602.

on Fourier theory, which encompassed diffraction. He was also famous for
monologous conversations. "There was then and is now no subject which
can be brought up on which Wiener does not have something interesting to
say. And with him the subject is always changing. I estimate I got in about
one question on Fourier theory per two or three hours of conversation, but
the answers were usually pay dirt."[8] Patterson was searching for a new route
around the phase problem. "I very soon learned from him the fact that I had
to work with the *Faltung* [correlation function], but it took me more than a
year to catch on to what it was all about."

The route you are looking for, Wiener told him, is a corollary of the
Wiener-Khinchin theorem. Fourier transformation, the mathematical relationship between the crystal structure and the diffracted waves, is also the
relation between the x-ray intensities and a certain contour map. This isn't
a map of the crystal structure; it doesn't show you where the atoms are. It
shows the relative positions of *pairs* of atoms instead. You still have to find
the positions of the atoms themselves, but that is easier than what you've
been doing.

Protein crystallographers seized on this alternative. They called the contour map a Patterson map, or a Patterson synthesis.

Patterson maps were complicated and contentious. What, in fact, could they actually reveal about a crystal structure? If essential information was missing from the diffraction intensities, Fourier transformation would not restore it. And even if the Patterson map did hold all the necessary information, how do you get it out? How does the contour map yield up the crystal structure? And mightn't different structures have the same map? Patterson addressed these matters in his landmark papers. The information you need is there, he said. (He was right, though he couldn't prove it.) The structure, when you find it, is unique. (He was wrong, and Linus Pauling promptly told him so.[9]) As for using the map to solve the structure, well, no, he had no method. Some trial-and-error guessing is probably necessary, he said, but less guessing than before.

Delta studied Crowfoot's second insulin paper and fired off a response. The Patterson map of a theoretical cyclol-insulin crystal matched the new data, she claimed. But her calculations depended on how the cyclols were arranged and tilted, and which points on them were assumed to scatter the x-rays. Hodgkin and her supporters argued that Delta's tilts and assumptions were arbitrary. Delta and Langmuir argued back. Denis Riley, Crowfoot's student, and Isidor Fankuchen, a young American in Bragg's lab (called then and ever after simply "Fan"), redid Delta's calculations and announced that she'd done them wrong. No, she retorted, you've tilted the cyclol a few degrees too many. The battle played out in the pages of *Nature*. They said, she said, they said, she said…and meanwhile the enthusiastic Niels Bohr made metal cyclol models in his lab.

This may be the only battle in all the worldwide history of science between two women with the same first name. One historian of Oxford science suggests they vied for the title of Protein Queen.[10] But their differences were more striking and more telling. Hodgkin (Crowfoot added her husband's name when she married in 1937) was the anti-Wrinch. Quietly brilliant, self-effacing, she stayed in the background, working in her laboratory, offending no one. She grew famous anyway and prizes rained down on her, especially after the Nobel. Never, ever, she insisted all her life, had being female been an obstacle to scientific success.

Their proxies, Bernal and Langmuir, met in person to thrash insulin out. Langmuir talked for hours. Bernal didn't budge. Nor did he budge Langmuir. So Bernal went public; he didn't want to, he said, but he had to.

He had to, for several reasons. He had rejected the cyclol model by this time on scientific grounds. His close personal ties to the other Dorothy must have weighed in his decision. But his greatest, and his stated, motive was to protect the fragile field of protein x-ray crystallography. It had yet to score a victory. Not one protein structure had been solved: not hemoglobin, Dr. Watson, not myoglobin, not insulin. Many scientists dismissed the whole effort as hopeless. Delta's erroneous claims would give the field an even worse name, discrediting all their patient work.

Bernal enlisted the "father" of x-ray diffraction in the cause. WLB was cautious. He had read the Wrinch-Langmuir article on insulin hastily, he said, because he could only borrow the journal from the library overnight. He refused to be drawn into the debate over cyclols. But, as the founder of the entire field of crystal structure analysis, he could, and would, assess Wrinch's methods. For Delta now claimed to have discovered a simplified Patterson map and a new algorithm for solving the structure from it. This was nonsense and he would say so.[11]

"W. reports," Warren Weaver noted after a phone call from Delta in the summer of 1938, "that not long before sailing for the United States [that summer] she became interested in the interpretation of x-ray data, certain work in this field having been quoted as giving rather strong evidence against the cyclol hypothesis.... Her study of existing methods led her to a new and, according to her, more powerful method, and using this method she soon obtained information which strongly confirmed the cyclol hypothesis."[12]

I speak not of peaks in contour maps, but of points in space, she said. Points instead of peaks, and points instead of atoms. Structure solving is a geometry problem, deriving one set of points from another.

A century earlier, Delafosse had replaced Haüy's fictive building blocks with points, a fruitful move that led to lattices and, eventually, to the discovery of x-ray diffraction. But points instead of peaks in the Patterson? "The substitution," said Bernal, "of points for peaks in the Patterson vector map, and the subsequent attempt to find a point system to which this vector map corresponds, is not a reliable procedure in the analysis of complex crystals."[13]

Had he said *feasible*, he would have been right, in those days before computers. But Delta's method *was* reliable in principle.

With peaks replaced by points, said Delta, finding the original structure—with points standing in for its atoms—is just a connect-the-dots problem. Suppose, for example, that your Patterson is the set of points (a) on the left in figure 10.

First, connect the dots to the center point, as in (b), to form a star (all Patterson maps have a center point). She called the star a vector diagram because each ray of the star is the vector between two points of the set.

Next, find a set of points for which the rays of the star are the vectors that join them (c). This is where her algorithm comes in: one doesn't just play

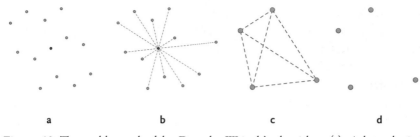

a   b   c   d

Figure 10 The problem solved by Dorothy Wrinch's algorithm. (a): A hypothetical point set Patterson map. (b): The same map, with the points joined to the center to form a star. The algorithm finds a second set of points for which the vectors between points are the arms of the star. (c): The solution for this particular example. (d): Same, with the vector lines removed.

with the rays, push them around, until they lock together; instead, she gave a straightforward procedure for finding a correct arrangement.[14]

Finally, erase the vectors, and you have the point set you were looking for (d).

Applied to crystallography, the algorithm starts with the Patterson map (points replacing peaks) and ends with the atomic distribution. (One, or more than one, but it finds them all.) It works for any number of points, she said, finite or infinite, and in three dimensions too. The procedure is methodical. A few technicalities may have to be finessed: you might hit a dead end, then you'd have to backtrack. But you will always arrive at the solutions. *Any distribution can be computed from its star.*

Months before Pauling alerted Patterson to the uniqueness question, Delta had seen into it, if not all the way through it. "Crystallographers assert that within the limits of inaccuracy of vector diagrams obtained from x-ray photographs, it is quite possible for two different atomic structures to give the same vector diagrams," she wrote to Langmuir.[15] "No doubt this is correct. On the other hand, the only point which interests me is—can two atomic distributions give the same vector map, a purely mathematical question [which] is surely capable of being settled." The vector map represents "a set of structures which are in some relation to one another; they must be transforms of one another in some way."

And now, back to the fray.

Delta had, in principle, solved the problem Wiener bequeathed Patterson: find a path from the map to the crystal structure. But the actual computations were formidable for even the simplest cases. Her critics, focused on insulin crystals, ignored her core idea. "I have now reread the article and letter published in *Nature* last Fall, and I have not been able to find anything beyond what is contained or obviously implied in the work of other x-ray men," Linus Pauling wrote to his former student David Harker. Harker, by slicing through the contours of Patterson maps, had made the theory somewhat simpler. "Have I overlooked some important idea in the mass of verbiage?"[16] Harker had missed it too. "Wrinch's insulin model agreed well with Crowfoot's data," he was happy to report. As for the method, he saw only that they had made a technical improvement in his.

Bragg weighed in to protect his field. "The 'geometrical method' proposed by Wrinch...is essentially the same as that used in these and other investigations," he concluded,[17] citing five papers. "Exaggerated claims as to the novelty of the geometrical method of approach and the certainty with which a proposed detailed model is confirmed are only too likely, at this stage, to bring discredit upon the patient work which has placed the analysis of simpler structures upon a sure foundation." In a companion paper, Bernal added. "What is new, however, is the use of certain additional assumptions and their application to cases where they are of extremely dubious validity."

Fast-forward 15 years.

"Like many other crystallographers who have discovered new methods of structure analysis, Wrinch optimistically considered the applicability of her method to the determination of the structures of crystalline proteins," wrote Lipson and Cochran in 1953. Henry Lipson was the author of a paper Bragg had cited in his put-down. "In the ensuing controversy, Wrinch's important contribution...was lost sight of, and little attention was paid to it until parallel results were obtained ten years later...by Buerger."[18] Martin Buerger, a crystallographer at MIT, known for his invention of an x-ray camera and for his work on crystal growth and twinning, built a new phase of his career on her insights.

"Dr. Wrinch's work...has been subjected to harsh criticism in the past, particularly by Professors Pauling, Bernal, and the writer." The writer is Fan—Isidor Fankuchen; the letter a review of a grant application for the National Science Foundation, the year 1958. "I can certainly testify, and willingly do so, that the criticism by J. P. [*sic*] Riley and the writer (published about twenty years ago) was based on an incorrect evaluation of the then available evidence."[19] By that time, Delta's algorithm—elaborated by others—was widely used in solving crystal structures simpler than proteins.[20]

Bragg recanted implicitly, using one of her drawings in his lectures in the 1950s and in his last, posthumous, book.

Dorothy Wrinch died before I went to Moscow; she never knew about my conversation there with Nikolai Vasilevitch Belov, Academician, past president of the International Union of Crystallography, Honorary member of the Mineralogical Societies of the USSR, USA, Great Britain, and France, Hero of Socialist Labor, holder of four Orders of Lenin; and recipient of the Lomonosov Gold Medal, the Soviet Academy's highest award.

I had been corresponding with N. N. Sheftal', a crystal growth expert at the Shubnikov Institute for Crystallography, and I wanted to work with him. An exchange program between the United States National Academy of Sciences and the Academy of Sciences of the USSR gave me the chance. Each side had a fixed number of months to parcel out to its scholars; I was given 11 and a plain but sunny apartment on the 14th floor of a residential building near the institute. My daughters Diana and Jenna, then 14 and 11, attended Russian public schools a few blocks away, bravely immersing themselves in a second new language and culture in four years. They dressed like their classmates in sturdy brown uniforms with black aprons and detachable white collars. Each week I snipped off those damned collars, scrubbed them white again, and sewed them back on. That winter was the coldest on record, dropping to minus 45 degrees, the crossing point of the Farenheit and Centigrade scales. Jenna kept her head and face tightly wrapped in a long wool shawl; we had to lead her around. Social isolation exacerbated personal weaknesses and professional rivalries; in the decade before glasnost, Moscow was a divorce factory for American residents.

But the lab was warm and relaxed.

N. V. Belov was the institute's superstar. Though nearly 90, he came to work every day, always wearing a black leather jacket, and sprinted the five flights of stone stairs to his office. He attended every seminar and asked sharp questions. Recently, I found the diary I'd kept sporadically that year. Belov had called me, the American visitor, in for a chat. "We consider Dorothy Wrinch the greatest American crystallographer," he told me out of the blue. "Not Martin Buerger."

Belov didn't explain what he meant, but a recent paper by one of his former students does. Belov had pointed out, back in the 1960s, that the first question about the Patterson map—do the diffraction photographs contain enough information to solve the structure?—had never been defnitively decided. Belov "suggested the idea of trying to find a solution [to the whole problem of structure analysis] by determining the full set of interatomic distances [from diffraction measurements], because the problem of finding the atomic map from this full set had been solved by Wrinch."[21] Now, 40 years later, Belov's student had proved it could be done.

And her algorithm was not just about x-ray diffraction. "Inverse problems" arise in image formation, pattern recognition, psychology of vision, error correction codes, DNA sequencing, and even further afield.[22] The algorithms are written for computers now. Their geometry is invisible, and they go by many names. "Wrinch" is not among them.

19 STRUCTURE FACTORS

The Lady Carlisle Research Fellowship was Somerville College's most prestigious. Under the terms set out in 1913, competition for the five-year award was open "to unmarried European women, British women, and Jewesses"; four of the five years must be literally spent in residence. In the spring of 1939, Delta applied for the fellowship to continue her research in "The Application of Mathematical Analysis to Biological Problems." D'Arcy Thompson wrote a discerning letter in support:

> One can hardly pass a fair judgment on D. W. without knowing her sad history. You doubtless know that she made an unhappy, even a tragic, marriage; from the consequences of which she still suffers acutely. A certain excitability, a certain forced gaiety, which one sees in her and which is apt to jar on one, all comes of frayed nerves, under circumstances which few women would have come through unharmed. What she wants is the chance of settling down to work, to work of a high class, without worry and anxiety for the next few years. I think it well worth while to help her to do so.

The committee voted unanimously to appoint her.[1]

But Delta was hedging her bets. "W. states that she is very anxious to move to the U.S. and find a position here," Weaver had noted the year before.[2] And she had told Lynda Grier at Lady Margaret Hall that she would go to America if war broke out.[3]

When Hitler's army crossed into Poland and Britain declared war on Germany, the Royal Society compiled a list of scientists willing to apply their expertise to defeat him. Delta signed up. Like Bertrand Russell and many other World War I pacifists, she supported this war (Eric Neville did not). But the Royal Society saw no way to use her services at that time.

Then how can I help? Oxford authorities put the matter to the university's controversial chancellor, Lord Halifax, who was also Britain's foreign minister. "H. wrote a letter saying that he would make suitable inquiries but that his offhand judgment was that W. would be most useful if she continued her research and lecturing 'in the United States,'" Weaver noted.[4] "After a short interval, during which H. presumably carried out the inquiries indicated, he reaffirmed this judgment. On the basis of this letter from H., W. was able promptly to secure an exit passport and special permission from the Treasury for the external transfer of funds."

Delta snatched Pamela from wherever she'd parked her and boarded the SS *Washington* bound for New York. Pamela wept bitterly, sure that the move was a ruse to get rid of her guinea pigs.[5]

"Going unheralded to Baltimore," Weaver continued, Delta "received a cordial invitation from Andrews [the chairman of John Hopkins' Chemistry Department] to spend the winter with his group. They have no funds to offer, but can give her a suitable room and facilities. Her stipend from Oxford will continue."

Four chemists at Johns Hopkins were working on proteins, the *Baltimore Sun* took note. David Harker was one of them. Inviting Delta had been his idea. He had never thought about proteins until "Langmuir decided that I was the person to talk to a Dr. Dorothy Wrinch who was, I believe, here in 1939, early spring, consulting, giving lectures. He invited me up from Baltimore to Jones Beach, and himself from Schenectady along with Dorothy Wrinch, and we had a lunch in bathing suits on Jones Beach." The picnic changed his career.[6] "And there was Dorothy Wrinch sitting with a leather attache case full of diagrams and little models, and Langmuir was talking madly about surface films of barium stearate and calcium stearate. I had an awful time diverting him. Finally, I said, 'Dr. Langmuir, I am fascinated by these surface films, and I would like to know more, but here you have brought Dorothy Wrinch to talk to me about the structure of proteins, and what do you think we should do?' So he said, 'All right, let's talk about proteins.' He was very definite."

"She taught a very nice course on the mathematics connected with organic structural chemistry," Harker remembered, "and she, of course, told us all about her 'cyclols,' as she called her models for protein molecules....After a while she wangled an invitation from Eleanor Roosevelt to lunch at the White House in favor of rights for women of some kind, and she described her lunch at the White House, and how they were all seated, and Mrs. Roosevelt came in and said, 'I'm so sorry, ladies, but Franklin is too busy to come to lunch today. Will you please accept his apologies?'...Sometimes she played the piano. She could read music very rapidly and very athletically. She was very gay and energetic."

The cyclol controversy raged on; her model was still, as Delta liked to put it, *sub judice*. "I wrote a paper with her on bond energies," Harker added, "which showed, as of that date, that it was impossible to determine on the basis of energetics whether the cyclol or the linear structure for polypeptides was the correct one."[7]

"This Pauling business gets me down," Delta wailed to Neville by slow—very slow—boat mail.

> He is a most dangerous fellow. He wrote a perfectly vile letter to D. Harker after our little article came out. Apparently H was quite ill for a week after getting it. They are as I have told you before desperately poor with his and her mothers to look after. As D told me about it I understood that he considered that the letter warned him not to expect support from LP in getting a better job, in fact that he was disinherited. I haven't seen the letter yet nor his reply in which, he says, he told him what was what, including remarks about his behavior to me which said H he had been reluctant to believe in before....
>
> I don't mind telling you Erice, that often I would like to quit. The fact that I can't makes me hold on of course, but if only I had an opportunity to do so I easily might. I get absolutely in despair, for I see the whole setup as a power syndicate just like Hitler's and only the strong and powerful survive...Even decent people hesitate to stand up to LP. He is bright and quick and merciless in repartee when he likes and I think people just are afraid of him. It takes poor Delta to point out where he is wrong: truly none of them would under any circs. The big paper on bond strengths and lengths has come back from JCPhys. with reports from six referees. They all seize on my comments which apply to P and want them deleted. They are cowards. Oh dear, why is it my fate in this field to have to point out the idiocies of the great?[8]

In Baltimore, Delta and Pam lived with a family with "3 spoiled kids, they make the most fiendish row all day and all night. I can't work." Pam had nightmares about Nazis and guns. "Darling Pam is growing so sweetly but she has the awful craving for comfort which is being satisfied partially at the moment by passionate devotion to the Church of all things—but I welcome it as she needs more comfort than I can give her."

Delta's forbearance didn't last. "We had a horrid row last night, my fault really, but I wanted to vent my feelings on the poor mite and found myself tearing into little bits with unbelievable venom her much prized horrid little calendar showing all the lent services and all the nonsense of her church. This made it clear to me how much I really mind.... Here was a large and complicated and very well organised institution which officially exists to give her and others comfort, so I rejoiced about her confirmation and church going. But Lord it hurts me frightfully. If I tell her one percent of what I think about the Ch. it may estrange us forever. So what to do."[9]

In 1940, Delta resigned her Somerville research fellowship: she would not bring Pam back to wartime England. Johns Hopkins gave her another unpaid year. The Rockefeller Foundation gave her a one-year extension but, they told her, it would be her last.

"The awful loneliness shown by: waiting about after lecs. Coming to CSH [Cold Spring Harbor] and WH [Woods Hole] so there shall be human beings to talk to. Attempts at friendship with people I don't like...Awful anxiety about the future of the work or even of us...Can't make the grade with people I have set up as my standard."[10] Neville, she realized, would never leave his wife.

In the summer of 1940, Delta met an old friend at Woods Hole. Reginald Ruggles Gates, F.R.S., a Canadian-born botanist and geneticist, had been a colleague of John Nicholson at King's College, London. Gates was still on the faculty but had decided to wait out the war nearer home. "Poor Gates has just finished getting his divorce," Delta told Neville.[11] "I foresaw all this when he went with us to South Africa in 1929 with his blushing bride.... He is lecturing at Berkeley, California.... His new book on Botany is now being gotten ready for publication and it sounds quite good, since he has managed to put in a bit of chemistry here and there, to say nothing of cyclols." She did not mention that they were talking of marriage.

In cryptic notes to herself, Delta listed the pros and cons that life with Gates might bring. "Sheer loneliness avoided, help in home, financial gain in BLG in Western hemisphere, moderation and self control, after the lecture one can go home, holidays and recuperation in non work periods, home shelter for Pam." She weighed this against her "dislike of companionship day and night, restraint of my actions, year over and no job? I can't be a LU [London University] wife. Somerville and Oxford in general. Certainly lost!...LMH [Lady Margaret Hall] my God." Gates told her outright: he would tolerate no philandering. Delta agreed to behave. "He is conventional (ethical princ. which will take toll) and I think this is best (contra Berties ideas.)"

Margery Fry advised her from London:

I don't think he [Neville]—in any generosity—could ask you to wait for him indefinitely, and I'm old fashioned enough to respect him for not throwing over altogether his responsibility to an unloved wife—it's one of these situations which don't seem to have got a "right" solution. Still, he can't hold you bound and himself not come to you, unless you've made a more definite pact with him than I think from what you say you have. In the abstract it would, I'm sure, be infinitely better for you (and probably for Pam) that you should marry again. *On balance*, that is: I suspect you'll find that though you'll lose your loneliness it'll involve—even if it's a reasonably happy affair, rather more give and less take than it looks like at present. As to your Canadian botanist, I don't think most of what you say—neither brains nor Stopes—matters much. But he's not *rigid*, is he? That would be a real difficulty, you'd bruise yourself against it, and it would cut you off from the other friends you do care for.[12]

That Delta considered marrying a man with whom she would so surely clash plumbs the depth of her misery. Like his first wife, Marie Stopes, the birth control crusader, Gates was a eugenicist and a racist. But, his biographer stressed, there was "no evidence that he was ever called a Nazi or a Fascist by his opponents…He was a loyalist, who still looked back with sorrow to the American revolution, a monarchist, a traditionalist, a loyal Canadian and British subject, an admirer of the achievements of the British Empire. He was a practising member of the Church of England."[13] Fortunately for everyone, the Gates-Wrinch relationship went no further.

American mathematicians, including the eminent, were switching to war work. Should I? Delta asked the Rockefeller Foundation. They raised no objection, but Neville did. "Many thanks for the remarks about Ballistics," she replied to a letter she didn't save. "It always surprises me the kind of turn you give to your remarks. I should have thought the only trouble would be if the work were mathematically insuccessful [*sic*], but here you are assuming that it would be perfectly simple for me to do, but that it would not be good to do it from the chemical point of view. I don't think that additional mathematical reputation makes it any easier for the chemists to kill me; rather the contrary. However I will bear this in mind."[14] In the end, she decided not to work on weapons. Not out of residual pacifism, nor with an eye to her future, but "to keep the torch of pure science lit" in a time of stress.

Delta cast about for an academic job. The Rockefeller Foundation refused to help her.[15] Karl Menger, then at Notre Dame, agreed to keep a lookout. "If

I should hear of any opening for a lady-instructor in mathematics I certainly shall think of you…Did you write to [Solomon] Lefschetz [at Princeton]? I think he is a great admirer of you and may hear of openings in the numerous Eastern institutions for women with which I have no contact."[16]

Otto Glaser, a biologist at Amherst College who admired Delta's work, tried to place her in Princeton too: not at the university but at the Institute for Advanced Study. He would, he told "Dear Doctor Dorothy Wrinch," write to his friend Walter Stewart, an IAS trustee who had taught at Amherst in Glaser's first years there; "in the early days of the Renaissance here we fought side by side."[17] Stewart talked with "certain members of the math group." This is not the place for her, they told him, though "they know of Dr. W's work and have great respect for it."[18] That was not the last word on the subject; in the spring Delta complained to Neville:

> Partly the reason for my despair is my visit to Princeton this week. Einstein on March 8 practically said he would fix me up you know. Said most extravagant things. Well I waited hour by hour for 18 days, finally telephoned and went to see him. He then said Veblen and others had taken a firm line, said their little bit of money was for maths alone and no nonsense about pp. They were I take it very fierce. Well E prefers to live in peace with them than to help me of course. He took the proposal to take me on to the [??] who is a little pip squeak. He said, new funds would be nice. E said well lets get them. He said well maybe we can maybe we can't. E said this fellow is a diplomat and never says no. But I think and E thinks I believe that he will not try to get funds.[19]

But Glaser had a plan B. Why not move here? It's a very nice place with two women's colleges nearby. You could influence a rising generation of scientists. He had already broached the idea to Amherst College's president, Stanley King.

Glaser was a biologist after D'Arcy Thompson's own heart. Joseph Needham had cited his work in *Order and Life*, and Glaser was on Weaver's must-meet list of scientists for Delta's first trip to America. She didn't get to Amherst in 1936, but she visited a year or two later and they bonded over polyhedra. With George Child, a junior Amherst colleague, Glaser modeled the growth and form of living cells by stacking minute plaster truncated octahedra, as if cells were built like crystals. Now, in 1940, he was using the same building block to simulate the growth and form of the tomato bushy stunt virus. Delta saved his virus models and later gave them to me; they're next to her cyclol model on my shelf.

Dorothy Wrinch in the "Smith Mobile" with unidentified men. Photographer unknown, copyright Smith College. Dorothy Wrinch Papers, Smith College Archives.

Glaser proposed to King that Delta be appointed visiting professor at Amherst, Smith, and Mount Holyoke colleges for one year. King consulted the other presidents and they all agreed; it would be an experiment in inter-college cooperation. Delta accepted with pleasure.

She was sensitive to local conditions. "I think it of the utmost impor-tance that everyone involved should take as much part as he will in making plans," she told Glaser. "I am notoriously poor at intuition about people, but I have a feeling I should go carefully in regard to chemistry at MH. EC is a very remarkable woman, does as good a job in her department as many more eminent people. I greatly wish NOT to tread on her toes."[20] EC was Emma Perry Carr, a well-known chemist; Mount Holyoke's Carr Laboratory is named for her.

At Smith Delta was welcomed by Gladys Anslow, a nationally known and widely respected physicist. She visited in April and gave a lunch lecture to the Smith College chapter of Sigma Xi; the title was "The Megamolecule." She enrolled Pamela in the Burnham School for Girls in Northampton and looked for an apartment nearby.

"I do rejoice with you over your good news," wrote Neville. "It is excellent to have a definite footing in the educational world, especially one that doesn't implicate you in the examining and administrative machinery. I take it you teach what you want, and have nothing to do with other people's wants."[21]

Today Five Colleges, Inc., the consortium of Amherst, Smith, Mount Holyoke, and Hampshire colleges and the University of Massachusetts, is a national model for intercollegiate cooperation. Five Colleges is not quite Oxford or Cambridge; the distances in the Connecticut Valley are a little too great and entrenched turf pride much too fierce. But cooperation has grown steadily. There is a Five College lecture fund; an intercollege faculty exchange; student interchange; Five College faculty seminars; joint faculty appointments; and even, in some fields, joint departments, majors, and programs. There is a common online library catalogue and rapid interlibrary loan, for which I am daily grateful, and a free and frequent bus service. In 1940, Hampshire College had not been imagined. The University of Massachusetts (then called Mass Aggie), Amherst, Smith, and Mount Holyoke had no formal connections, though a chemistry lecture series had been running for a decade and the Connecticut Valley Mathematics Colloquium had begun a year earlier. Delta's joint appointment was the first in the Valley's history.

Delta and Pamela did not move to Northampton after all. They moved to Otto Glaser's home in Amherst instead, a large three-story white frame faculty residence at 233 South Pleasant Street, a short walk from the home where Emily Dickinson lived and died. F. B. Hanson, a Rockefeller official, noted in his diary in August 1941, in Woods Hole, "G. married Dorothy Wrinch in the office of the Director of the MBL on the day after W completed her work under NS grant."[22]

"Dearest Dorothy," wrote Margery Fry in September. "Can a bombshell be an agreeable thing? Anyhow *your* news is a good bomb! I'm *very* glad about it...I'll write to Eric."[23]

Delta wrote to him herself a few days later.[24] The decision had been sudden. On July 27, Otto "spoke for the first time of something other than molecules and geometry. I was extremely amazed and surprised and taken aback....I gave the matter much thought during the night and for days after. But the wonderful thing happened. Within a week I was quite clear about it too."

A *permis de marriage* was required, three in this case. The college presidents gave them their blessings.

"There then remained about 6 weeks before term. Pam was in Maine so we went off there. Pam, the darling child, telegraphed and wrote many times in the course of about ten days pointing out that I had made one awful marriage and that I ought to think carefully. However when we got there, she hardly recognised her completely transformed mother and sitting back in the grass said, well, I guess it is settled....Erice dear, I am happy at last. I take it that our friendship goes on?"

"God bless thee," wrote D'Arcy Thompson, "—and he will bless thee!—as I once heard a great man say in much the same circumstances. I never met your Otto, but I've known his work for many years. He has been generous in sending me his papers, and they have a box to themselves—you only have half-a-box—in my room at College."[25]

"It's good about Dot's marriage," Dorothy Hodgkin wrote to Fan. "I've been hoping something of that kind would happen for her sake and Pamela's. Wad says Glaser is very nice and suitable."[26]

"The bride," noted the *New York Times*, "a Rockefeller research fellow and a member of the chemistry faculty at Johns Hopkins University, recently was appointed Chemistry Professor at Smith, Mount Holyoke and Amherst Colleges. Her work on protein molecules started a controversy among the leading chemists, physicists and biologists of this country and Europe."[27]

Each week Smith students filed into John M. Greene Hall for compulsory chapel. Smith was not an infidel place, but the chapel sermons were food more for thought than the soul. Dorothy—she dropped nicknames now and ever after—laid out her vision before the 3,000 young women neatly dressed in skirts and sweaters.

> The new synthesis now emerging consists in the piecing together into one consistent whole of facts and findings from all the many different fields of work and working them into a backcloth of geometrical ideas....As Newton said in the Introduction to his *Principia*, "Gloriatur Geometria quod tam paucis principiis, tam multa praestet." It is the glory of geometry that from so few principles so much can be derived.

Her course on molecular biology, "the first of its kind to be given in any center of higher education," was a crossdisciplinary reading seminar. Years of dashing between Cambridge and London, looping through Somerville, Lady Margaret Hall, and St. Hilda's, had trained her to run this maze. She taught at each college each week, crossing the Connecticut River to Smith and the Holyoke Range to Mount Holyoke. She tailored her course to the students on each campus and to the interests of their faculty, and students and faculty gave research presentations. Delta also gave three lectures for the general public: "Patterns in Biology" at Mount Holyoke, "Patterns in Chemistry" at Smith, and "Patterns in Medicine" at Amherst. Elizabeth Horner, a biologist at Smith, told me she attended one of them. "She was flamboyant, theatrical. It was fun. And she had a wonderful accent. But we weren't convinced she was right."

At Smith, the course met on Wednesdays. The first session, says the syllabus, was devoted to "the lattice of diamond and graphite," the next to "Crystal structure of straight chain paraffins." The third, October 8, was Mountain Day, for which the course was canceled. The last two lectures were on "the fabric theory of the structure of native proteins" and "the native proteins in their biological setting." The final exam was an essay.

"It was striking to see what excellent talks resulted," Dorothy reported to Smith's President Davis in May.[28] "They served a useful function in giving the students experience and confidence in exposition. They were also quite remarkable in the way in which they made the various students feel that they were collaborating, and would perhaps be able to collaborate further in the future.

The experiment was successful, Amherst's President King told his trustees. Yet he would not renew her appointment, and she knew it early on. "My future, in the academic line, is quite difficult," she told Neville three weeks after Pearl Harbor. "I don't see how Smith can use me in perpetuity, tho I wish to heaven they could: nor Holyoke either. This in spite of really very great success at Smith. Amherst of course won't continue, spec. under the circs and boys college, etc, but things have been going very nicely there, with the chemists, who are an interesting set, gradually seeing the point."[29]

"The circs" alone—America's entry into the war—would have ruled out further intercollege experimenting. Enrollment at Amherst plummeted to 90 as students enlisted. A quarter of the faculty left for war work, and many of those remaining on campus put aside their usual activities to teach a large contingent of officers in training. Smith and Mount Holyoke refashioned themselves too, training officers for the Waves (Women's Auxiliary Volunteer Emergency Service). "U.S. Navy Waves billeted at Smith College were marched down Main Street three times a day for their meals at Wiggins Tavern," says the Tavern's current website, quoting one of them: "there were two great things about being stationed in Northampton: 'You survived the war, and you got to eat at Wiggins.'"

But Smith reappointed her, as an honorary professor of physics. It is not clear what her duties were or whether she was paid, but for the next 30 years her Smith colleagues considered her, and she considered herself, some sort of member of that department.

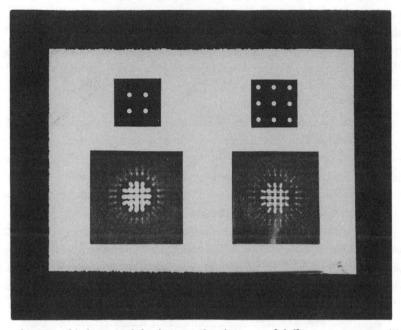

Dorothy Wrinch's lantern slide showing the elements of diffraction geometry. Top, two small "masks"—points punched in metal plates. Bottom, the diffraction patterns produced by these masks.

Amherst College held a summer session in 1942, the first in its history. Though he was a trustee of the Marine Biological Laboratory, Otto Glaser stayed in Amherst that summer and visited Dorothy and Pamela in Woods Hole on weekends. He wrote chatty letters every day, "I love you ∞" (∞ is the mathematical symbol for infinity). In lieu of his signature, Otto sketched his cuboctahedral building block.

That summer an old acquaintance of Otto's, Laurence Gross, returned to Amherst after 12 years in Boston. Gross was a jack-of-many-trades— woodwork, metalwork, painting. "He is intelligent and resourceful and has wonderful fingers," Otto told Dorothy.[30] "Some time not too far off Meyer [Otto's technical assistant] will retire and Gross I think would be an excellent person to put in his place...I'm wondering if I should try (which would not be difficult) to make him 'model conscious'? Do you think that between us we could keep him busy?...If we could use such a person for the purely scientific aspects it might be well to see if we could get a little grant for that purpose. All this is a bit nebulous at the moment."

Inspired perhaps by Otto's virus models, Dorothy was thinking of building blocks too.

What are the building blocks of a diffraction pattern? How are complicated diffraction patterns made up of simple ones?

In July 1943, Dorothy sent an inscrutable manuscript on diffraction building blocks—"structure factors"—to Martin Buerger at MIT; he edited several journals. What's this about, he asked her. In a long letter, she explained. "I thought your letter explaining the paper was a masterpiece of scientific philosophy," he replied. "I am sorry that you do not introduce the paper with these sentiments. The paper is a little cold, but the accompanying letter positively scintillates."[31]

Dorothy set about expanding her letter with computations, diagrams, and graphs. Mr. Gross would help with this; there were funds for an assistant in Otto's Rockefeller grant.

20   AMHERST COLLEGE WIFE

Otto Charles Glaser was born in Wiesbaden, Germany, in 1880; the family moved to Baltimore when he was small. After graduate school at Johns Hopkins he taught biology at the University of Michigan until 1918, when Alexander Meiklejohn, then president of Amherst College, persuaded him to join its faculty. Meiklejohn was shaking things up. He hired and fired, often bypassing protocol. He innovated and overspent. The trustees looked on with growing alarm. When, in 1923, they sacked him, Otto's friend Walter Stewart and eight other professors resigned in protest.[1]

With Meiklejohn gone, Amherst remained what it had been, a significant node in the interlocking web of America's elite, a web spanning day schools, prep schools, colleges, suburban enclaves, corporate boardrooms, and the corridors of political power. Its neighbor colleges, Smith and Mount Holyoke, strove to turn Amherst men's sisters into elegant, well-read prospective wives-to-be, schooled in gracious and thoughtful behavior.

Otto "was handsome, likable, and known favorably to many of the leaders in the field as an independent and promising thinker," says one obituary.[2] "He possessed an infectious enthusiasm for teaching biology to all undergraduates as an important educational experience, an essential aid in understanding their own limitations and capabilities in the twentieth century world. He was brought to Amherst by President Meiklejohn for precisely these reasons, and right up to the time of his death, his educational ideals and enthusiasm remained essentially unchanged." At Amherst, Otto rose to the top. His colleagues elected him to the all-powerful "Committee of Six" again and again. In 1939, President King appointed him Harkness Professor, making him the best-paid member of the faculty. In 1941, Glaser became acting president of the college, in charge whenever King was away.

"The Biology Department of Amherst College is, among institutions of that size and character, a remarkable one," a Rockefeller Foundation grant officer wrote in 1934.[3] "The two senior professors are both starred in American Men of Science. Professor Otto Glaser (head of the department) is a physiologist carrying on research in growth phenomena." Otto's brother Rudolf, an entomologist at the Rockefeller Institute for Medicine, sent him data and specimens. The other senior professor, H. H. Plough, the officer continued, "one of Morgan's pupils, has worked on the effects of temperature on crossing-over in the chromosomes of Drosophila and more recently on the effect of high temperatures in the producing of mutations...No other science department at Amherst presents a comparable picture...The officers view this as an attractive opportunity to create, strictly within program interests, an influential center of research in a comparatively small institution." The foundation's generous five-year grant to Amherst's biology department was earmarked for research in genetics, experimental embryology, and growth. In 1939 the grant was renewed for a second five years.

Dorothy later told her niece Talitha that Otto paid alimony to six previous wives, but he had been married only twice before. His first wife was also a Dorothy (Gibbs Merrylees); they married in 1909, had two children, and divorced in 1933 with permission from President King. In those days and long afterward, small-college presidents watched over every faculty household, the professor's effectiveness ever in mind. King did not hesitate to approve the divorce; he thought it would be good for the college. In King's opinion, Glaser's performance had weakened as his marriage collapsed.

While the first Mrs. Glaser was off getting the divorce, Otto met the second. This Mrs. Glaser, King noted pointedly, "devoted herself exclusively to her husband. Professor Glaser blossomed, did fine work, was recognized by the College and by his colleagues."[4] But she died of cancer in 1940.

Again Otto remarried within a year.

Alone, it seems, among the Amherst wives, the second Dorothy Glaser (and third Mrs. Otto) belonged to the wives' world and their husbands' world too. In the 1930s and the war years, refugee scholars and artists from Europe found temporary haven in the Connecticut Valley. When Dorothy Wrinch arrived in 1941 she joined acquaintances and friends of friends. Antoni Zygmund, the Polish mathematician Hardy so admired, was teaching at Mount Holyoke (in 1946 he moved to the University of Chicago, where, among other things, he taught me calculus). George Pólya, the famous Hungarian mathematician, spent the year 1941–42 at Smith before settling at Stanford. She reconnected with Hermann Muller, an expatriate American

Otto Charles Glaser and Dorothy Wrinch, 1946 or 1947. Photograph by Andrew Donald Booth.

geneticist who had been hired by the Amherst biology department on a temporary basis in 1940.

In 1912, the year the Meiklejohns came to Amherst, the wives of this all-male faculty started a club. "The Ladies of Amherst" met seven times in its first year: an informal reception, a play reading, a sewing session, a talk by Professor Crook, an informal tea for the faculty, a gabfest, and last, a business meeting. This set the pattern for the next 60 years.[5] The club made the wives of new faculty welcome and taught them the ropes, did charity work, held socials, and, once a year, listened to a lecture.

When Otto Glaser divorced his first wife, the secretary of the Ladies of Amherst crossed "Mrs. O. C. Glaser" off the membership list. It reappeared the next year without a hint that the bearer of this title had just acquired it. In 1941, again without a footnote, the label passed to Dorothy Wrinch. She not only joined the Ladies of Amherst, she stayed in it. She had blanched at the prospect of being an LU wife, but she became an AC one.

A constitution drawn up in 1938 specified the purpose of the Ladies of Amherst and the number of meetings (nine), the annual dues ($1.50), and the committees (social service, entertainment, decorating). Article II, the document emphasized, "has excluded discussion of political, religious, and controversial topics." The Ladies usually met at the president's house, but hostesses rotated. Instructions were precise: fourteen dozen sandwiches and the same number of cakes at each meeting, and two lemons, a quart of cream, a pound of coffee, and a quarter pound of tea. Sugar should be purchased in two-pound packages. The food and the table would be arranged by Mrs. King's servants.

Ten years later the Ladies found a club room, voted to forgo the food, and donated the money to war relief. But later the teacups returned.

"Even in the '60s we took the meetings very seriously," a friend told me. "We got dressed up, had our hair done, and had lunch at the President's house." The faculty was still small, less than 100; most lived in college-owned houses within walking distance of the campus and hearing distance of each other. This ensured a culture of subterranean jealousies and resentments. "You'll never get tenure here," the department chairman's wife and next-door neighbor hissed at my friend's husband. His two-year-old had thrown a tantrum, the neighbor had intervened, and he had told her to butt out. Though unstintingly praised and encouraged by his colleagues and the administration until then, he was not reappointed.

If "Dorothy and Linus" is an opera, "Dorothy and Otto" is a novel. Amherst College is also the setting for Alison Lurie's novel *Love and Friendship*. But that is a comedy; this novel is tragic.

To judge by the oral histories in the college archives, in the 1930s the Ladies, well-educated all, vied for supremacy in self-effacement. The first Mrs. Glaser had to be the center of attention, Mrs. Theodore Soller recalled with distaste, but the second, though so gorgeous she lit up the room, always turned the spotlight toward *him*.[6] The third Mrs. Glaser did not self-efface, but many Ladies were Smith graduates and would have respected her for teaching at their alma mater. Dorothy walked a tightrope through this clubby valley. And then she fell off.

All through the winter of 1943–44 she taught at Smith and expanded her paper on structure factors to book length; Martin Buerger would publish it in a new crystallography series. She also practiced the piano. On February 16, 1944, the Ladies held an evening meeting at President King's. The program was a concert, "Mr. Leonide Goldstein violin, accompanied by Mrs. Glaser on the piano." Goldstein, a hero of the French Resistance, was a professional musician and a pioneer in electroencephalography. His research overlapped Glaser's, Plough's, and Muller's; he had worked closely with Muller before joining the Resistance.

The Ladies of Amherst kept an album, but no photograph of this performance is in it. I wish one were; I would scan it, enlarge it, frame it, a memento of a high peak in the hilly contour map of Dorothy's life. The scandal broke three months later.

Helen Weaver donated her father's papers to the Rockefeller Foundation Archives, all but one slim folder. About 1960, apparently to prepare for his oral history interviews, Weaver asked longtime colleagues in the foundation world about grants to women back in the day. Helen sent me copies of his letter and the replies he received. And that is how I uncovered the scandal, so carefully buried for sixty-some years.

"It is my duty to report to you an incident which recently occurred at the College which is unpleasant and, I am happy to say, unique in my experience," President King told the Amherst trustees on May 29, 1944.[7] "The incident involves Professor Otto Glaser, our senior professor of Biology. Disciplinary action has been taken by the President, the facts have been reported to the Rockefeller Foundation, and the matter is now about to be closed." Closed for good: the Amherst trustees' records are permanently sealed. (I am grateful for permission to read and quote President King's report.)

The dean had blown the whistle: Glaser's assistant, paid with Rockefeller money earmarked for research in genetics, embryology, and growth, was in fact spending 80 percent of his time working for Dr. Wrinch.

"Professor Glaser had planned it this way," King explained. "On September 5, 1943, in offering a position to Mr. Grose [sic], he said, 'There will be a certain amount of photography—some line drawing—computations with the aid of tables and a comptometer—model making—etc.—all in connection with some joint work that my wife and I are doing.'"

The president gave the trustees a bit of background: the vicissitudes of Glaser's previous marriages, and Dorothy's one-year experimental joint appointment. When the experiment ended, she looked for research funds. Glaser had asked King whether he "would accept a subvention of perhaps $10,000 from the Merck Company, manufacturing chemists, to support the work of Dr. Wrinch at Amherst, with an undertaking of the part of Amherst to provide laboratory space but without other commitments from Amherst." King had turned them down. "I conferred with my scientific colleagues and declined the request. I suggested to Professor Glaser and Dr. Wrinch that they offer the idea either at Smith or Mount Holyoke. I pointed out to Professor Glaser and Dr. Wrinch that thereafter Dr. Wrinch's connection with Amherst would be only that of the wife of a professor. I regarded Dr. Wrinch's official connection with Amherst as concluded, with the exception of the fact that she was invited by the biologists to attend some of their seminars and she was invited by the Department of Physics to conduct one seminar for one session."

Why did King refuse the Merck money? He didn't say. Was space tight? Did he fear a wedge in the all-male citadel's door? Or was he concerned for the Amherst faculty family, in which a good wife "devoted herself exclusively to her husband"?

And now, King learned, the Glasers had gone behind his back. He took swift action.

"No matter how patient a president may be, the hour is bound to arrive, if he remains in office long enough, when he has to insist on a showdown with a disaffected faculty member," writes King's biographer.[8] "But the problem of removing a full professor, no matter how obviously unworthy or incompetent is not easy in a college like Amherst. The president always wishes, in such crises, to have the support of his faculty colleagues…in such unhappy situations King was diligent in assembling evidence, sensitive to currents of public opinion, and anxious to be just to everybody." The biographer did not name names or give dates.

In this particular case due diligence took less than two days.

The dean had informed King "about May 17." King "immediately conferred with Professor Glaser and found there was no question as to the facts. The following day I called a meeting of the department and summarily suspended Professor Glaser from the chairmanship. The department felt embarrassed to choose a new chairman and so I appointed myself chairman *pro tem*." (Two days later, on May 20, King appointed a member of the department Acting Chairman to serve until Professor Plough's return from military service.) Still on May 18, King wrote to Warren Weaver, saying he would be in New York on the 26th; could he meet with him about a matter related to the grant?

Weaver was away on war-related work; King met instead with his old friend Raymond Fosdick, president of the Rockefeller Foundation, and foundation officer F. B. Hanson. King offered to pay back the $580 Gross had received for work he wasn't supposed to do. ($580 in 1944 dollars is about $7,300 today.) The officers praised him for his skillful handing of the matter and told him to keep the money.

Concluding his report to the trustees, King put the blame where it was due. "It should be noted that the Rockefeller Foundation know Dr. Wrinch and her work. At some time in the past they provided a subvention to support her research but after investigation they decided not to renew it. Dr. Wrinch is well known by the scientists of at least three leading Eastern universities. The scientists at these universities have been predicting for three years that she would cause trouble at Amherst; she has."

The Rockefeller Foundation kept track of its pennies. Amherst's annual financial reports to the foundation list not only the amount spent on each and every item but also the company or store from which each item was bought. The smallest changes from the approved budget had to be approved in advance, and more than a few were denied. "There was a culture of accounting at the foundation," a Rockefeller archivist explained to me. "You have to remember that John D. Rockefeller started out as a book keeper. Standard Oil made its money on fractions of a cent per barrel." In his oral history Warren Weaver stressed the painstaking job of the foundation's comptroller: "He has to see to it that a philanthropic foundation maintains the most meticulous scrutiny of all its accounts, that it can always explain, down to the literal last penny, exactly what it has done with every cent, and that all of these expenditures have been legally proper, have always been backed up by the right set of actions by the right people."[9]

King ran Amherst College that way. After graduating from Amherst, he studied law, joined a shoe manufacturing company, rose to eastern manager and director of the International Shoe Company in Boston, and retired young. He was elected to the Amherst Board of Trustees in time to help fire President Meiklejohn for profligacy. King, the first president in the college's history with a business background, watched not only the bottom line but all the lines above it. The foundation's grant was an investment in Amherst education, vested in specific individuals with specific portfolios: genetics, embryology, and growth.

Glaser had not used his grant funds for the exact stated purpose. In King's eyes, the sin was mortal.

Even so, the vehemence of his reaction is bewildering. There may have been no question as to the facts—Gross *had* helped Dorothy—but their interpretation was less black-and-white. And stripping Otto of his department chairmanship was just the beginning. Nine months after his visit to the foundation on the matter, King met with Hanson again. "King refers to the fact that Child has been dismissed and that the Executive Committee of the faculty is considering Glaser's situation, looking toward recommendation for the date of his retirement," Hanson wrote in his diary. "In any event, the College will not put any more money into Glaser's research and will not request funds from outside sources for Glaser."[10]

George P. Child had come to Amherst in 1935 as a laboratory assistant and was quickly promoted to instructor. "Child was one of the best students of Prof. A. H. Sturtevant," a Rockefeller official noted. (The name Sturtevant spoke for itself, he didn't need to add "at the California Institute of Technology.") The official continued, "Curt Stern (Rochester) considers him a young man of great promise. FBH [F. B. Hanson] believes Child is Jewish, but says he makes an excellent impression. He proved so able and valuable

to the department that he was given a permanent post as instructor beginning in the fall of 1937."[11] Mrs. George Child made an excellent impression
too; in 1941, she was secretary-treasurer of The Ladies of Amherst. When the
United States entered the war, Child wrote to Weaver to ask how the department could serve. In 1943–44, he taught physics to officers in training. Yet,
Hanson noted in his diary the next fall after a conversation with Plough,[12]
"King thinks Child should be dropped with a year's full salary. He is not an
independent worker, does not get along with people. He is Assistant Professor,
but not on tenure. Child has too many interests, drops his own work to help
others, etc. P is not at all convinced that Child should be dropped, but at the
same time recognizes his shortcomings."

King dispatched Hermann Muller too. "It was P.'s hope when M went to
Amherst that a permanent post in their biology group might be found for
him. President King, on the other hand, thinks they should not keep Muller
for more than one more year after next June [1945]. King's criticism of M is
that he has not become known outside of Biology either in the College or in
the community. M apparently keeps very much to himself and continues to
turn out good research papers. Plough recognizes the force of King's opinions
about Muller, but believes they should find some way to continue his salary
on a year-to-year basis for a somewhat longer period."

Perhaps it was just coincidental that Professors Child and Muller were
Otto Glaser's friends.

Child resigned in the spring of 1945 and entered medical school at the
University of Georgia. Muller left in 1945 too, to take a research profes-
sorship at Indiana University. In 1946, he was awarded the Nobel Prize in
Medicine or Physiology for his work on x-rays and mutations.

The novelist who writes this story will conjure the aftermath: a department in turmoil, a stellar academic career in ruins, a president undone by his
outrage and his haste. King was, it seems, forced to retract one small measure. Acting Chairman Schotte told Hanson in the spring of 1945, "President
King has put at the disposal of the Department the necessary funds for the
maintenance of a research assistant to allow Professor Glaser to continue his
own research projects during the next year." Beneath Schotte's signature King
wrote, signed, and dated in a severely shaky hand, his agreement and support. He retired in 1946.

When Marshall Chadwell of the Rockefeller Foundation visited Amherst
in October 1947, he found Otto a broken man. His brother Rudolf had died
in September, an apparent suicide. "MC met G for the first time and chatted
with him on more or less general terms for about five minutes. G appears
worn out; complains that he has no assistants and is in semi-retirement."[13]

But the Executive Committee had not pushed him out. Instead, it voted in rules and procedures for retirement. Otto retired in 1948, after 30 years on the faculty. He died in 1951, of kidney disease and diabetes. Charles Cole, Amherst's president and King's successor, wrote to Dorothy, "I have always admired him so and have been so aware of his contributions to science and to Amherst College."

As ever, Dorothy persevered. As Lynda Grier once said, "Her power of putting through any work that she undertakes is remarkable; she would allow nothing to stand in the way of her carrying through whatever she put her mind to." Her book, *Fourier Transforms and Structure Factors*, was published in 1946. On November 19, 1947, she lectured to the Ladies of Amherst, in her own name, on "Science as an Art."

But foundations never forgot or forgave her. Indeed, the Wrinch-Glaser scandal ballooned in Warren Weaver's memory. He had not made many grants to women: they tended to get married and leave the profession, thus wasting the foundation's investment. In his letter, Weaver recounted the few he remembered, Barbara McClintock, Dorothy Hodgkin, and a botanist whose name escaped him. And this: "For a time I was conned into thinking—or hoping—that Dorothy Wrinch was going to be a major figure in connection with the basic problem of the structure of proteins. She was given some modest aid; and after she went to Amherst and was married to Otto Glaser. He received a grant, some of which was improperly turned over to her. This was so smelly a deed that Amherst refunded the entire grant to the RF. This is the only case of this sort that I can recall ever occurring in connection with an RF grant." Yes, said the few who replied to his letter, that's how we remember it too.[14]

The postwar era saw changes in science funding. The National Science Foundation was established. Rockefeller prewar practices—fund the man, not the project; gather opinions like bees in the meadow—were replaced, officially at least, by more formal, accountable, procedures: detailed proposals, letters of recommendation, and review panels with explicit criteria.

In 1948, Dorothy's Smith colleague Gladys Anslow received a grant from the Office of Naval Research to study the structure of hemoglobin and other biologically important molecules and "the flux which holds them together." She had written Dorothy into the grant. "Under the Navy contract," said the Smith College news release, "Dr. Wrinch will continue her mathematical analysis of X-ray studies of protein molecules with the help of various calculating devises recently developed at other institutions."[15]

Gladys helped Otto too. In 1949, they published a paper, cited many times since, on "Copper and Ascidian Metamorphosis."[16]

In Woods Hole, in the summers, Dorothy lectured on proteins. "It was a Friday evening seminar in the summer of 1949 when she showed her cyclol models," says Carolyn Cohen, now a leading biophysicist, an expert on the structure of muscle. "The beauty of these structures and the clear conception she had of the importance of proteins led me to appreciate, for the first time, the significance of this problem."[17]

But a whiff of "something not quite right"—as Smith colleagues put it to me—lingered on.

VI I DIED FOR BEAUTY

I died for beauty, but was scarce
Adjusted in the tomb,
When one who died for truth was lain
In an adjoining room.

He questioned softly why I failed?
"For beauty," I replied.
"And I for truth,—the two are one;
We brethren are," he said.

And so, as kinsmen met a night,
We talked between the rooms.
Until the moss had reached our lips,
And covered up our names.

—EMILY DICKINSON

THE SEQUEL

After the war, J. D. Bernal returned to his crystallography laboratory at Birkbeck College. Now the challenge was computing. No
calculator yet devised could handle the massive diffraction data
from proteins and other biological molecules. Electronic calcula-
tors developed for military purposes were being improved for civil-
ian use, but their cost kept pace with their speed. Bernal arranged
a Rockefeller grant for a younger Birkbeck colleague, Andrew
Donald Booth, to visit laboratories in the United States.

Ship passage was hard to come by in the wake of the war, but
Bernal got Booth on a "dry" ship carrying war brides and returning GIs. From Halifax, Booth proceeded to the Adirondacks to
address a crystallography meeting at Lake George. "At the lake-
side hotel where the conference was to be held, I was shown to
my room and started to put away luggage," he told me. "To my
surprise I opened a wardrobe and found it full of women's gar-
ments. I rang the Desk and received fulsome apologies and a
guarantee of action. A few minutes later a pleasant, middle-aged
English Lady arrived who introduced herself as Dr. Wrinch. I had
heard of Dorothy before 1946, largely from scuttlebutt from some
of Desmond Bernal's crystallographic staff, some of whom were
hostile to non-Socialists and given to unkind nicknames, such as
Rosie for Rosalind Franklin and Protein Dorothy for Dr. Wrinch.
I had not yet met her. This was the start of a long and fruitful
friendship."

Booth spent his two months in America on the road. In
Rochester, the chemist Maurice Huggins took him out on his sail-
boat as well as to the Kodak laboratory. He flew to California,
reaching Pasadena after many delays and layovers; he gave an
impromptu seminar at Cal Tech, which Pauling attended. A
few days later, he "talked protein both with L.P. and with one
Dr. Corey who is the right hand man." Booth told Dorothy, "I had

occasion to point out the extremely bad taste of the L.P. criticisms of the D.W. protein structure and took the opportunity of pointing out their invalidity. I also made some suggestions which were stonily received."[1] Returning to New York at night too late to find a hotel room, he camped out in Grand Central Station. At the end of his trip he relaxed with Dorothy and Otto in Woods Hole.

Booth was not, he told Warren Weaver at the end of his trip, overawed by the state of American computing. But his stop in Princeton was worth the whole journey. John Von Neumann was inarguably one of the greatest mathematicians of the twentieth century. Born in Hungary in 1903, he emigrated to the United States in 1930 and three years later, with Albert Einstein, was appointed to the faculty of the new Institute for Advanced Study. There he settled down, power ever-accruing, until 1957, when his life was cut short by cancer. The computer architecture he designed in 1945 for machines bigger than elephants is still the basis for computer design, though now they are smaller than an elephant's eye. "I had a long discussion with Dr. von Neumann and came to the conclusion that a machine built in accordance with some of his ideas would revolutionise crystallographic computing, and would be especially applicable to problems of biological interest," Booth told Weaver.[2]

"Bragg and Lonsdale are very impressed by your book [*Fourier Transforms and Structure Factors*]," Booth wrote to Dorothy, from England now. "Bernal also—so he told me, but I don't think he has written to thank you yet and most likely will forget altogether."[3]

Her monograph was not the student-friendly introduction that editor Buerger had hoped for, but it delighted the experts. Dorothy pointed crystallographers in a new direction. WLB had taught them to think lattices first, molecules second; she argued the reverse. The ABCs of diffraction patterns, she said, are diffraction images of small groups of atoms. Decipher these first, and then arrange them in the crystal.

A "short and excellent monograph," wrote William Lipscomb, a future Nobel laureate. "Dr. Wrinch has recorded the Fourier transforms of a large number of structural types, many of which occur frequently in a wide variety of crystals.... This excellent book can be recommended highly to anyone interested in the ways in which the Fourier transform can be used in structural analysis."[4] The monograph was "well received," historian David Berol concurs: "In 1952, John Kendrew, working with Max Perutz on the structure of hemoglobin, cited Wrinch's later work [i.e., the monograph] as one of the 'explicit foundations' of their most recent progress."[5]

Booth built a better machine, the Automatic Relay Computer, but this was a stopgap: it had paper input and output, but no memory. In 1947, again with Rockefeller support, he returned to Princeton for a six-month stint. This time Booth brought an assistant with him, a young mathematician named Kathleen Britten. (They married in 1950.) "Johnny von Neumann never prepared his lectures," Booth told me. "He would start on the left-hand board of the lecture room and, after a short time would say 'I should have proved the following theorem last time.' He would then start on another board. This would happen several times when his secretary would come in and say 'Washington on the phone professor,' at which point he would leave not to return. Many of the audience felt that this was prearranged."

Von Neumann's arc intersected Dorothy's through their interest in proteins. In a famous letter to Norbert Wiener (November 29, 1946), his equal on the genius scale, he sketched his reductionist reasoning:

> I would, however, put on "true" understanding the most stringent interpretation possible: that is, understanding the organism in the exacting sense in which one may want to understand a detailed drawing of a machine, i.e., finding out where every individual nut and bolt is located, etc.
>
> It seems to me that this is not at all hopeless…The major proteins have molecular weights 104–105 (the lowest one known actually appears to be only about 7,000) and the determination of authorities like Langmuir and Dorothy Wrinch consider it promising. Langmuir asserts that a 2–4 year effort with strong financial backing should break the back of the problem. His idea of an attack is: Very high precision x-ray analysis, Fourier transformation with very massive fast computing, in combination with various chemical substitution techniques to vary the x-ray pattern. I realize that this is in itself a big order.[6]

Two weeks later, after meeting with Wiener, von Neumann wrote to Dorothy, whom he had met before. He'd seen Langmuir recently, he told her, and had outlined his idea. The project would catapult the computations she'd done in *Fourier Transforms and Structure Factors* from small clusters of atoms to megamolecular protein molecules. "He showed a good deal of interest in this," von Neumann continued, "and he told me that D. Harker was actually working at the General Electric laboratories in Schenectady with parallel cm wave beams, with a definite idea towards x-ray crystallographical applications. We agreed that I would visit Harker at Schenectady sometime in January and discuss these things with him and Langmuir. I think that it would be most desirable, that you, too, should participate in these discussions."[7]

What Johnny wanted, Johnny got: the project was set to go. The RCA Laboratories in Princeton would supply instrumentation and personnel. "We will also have to provide crystallographical and chemical guidance and the actual models. All this would, of course, depend quite vitally on the possibility of a continuous cooperation with you. May we count on that? I think that I can make all the necessary arrangements for your visits, for expenses in connection with making models, etc."

"What I want is to transfer my work bodily to Princeton and to find some academic niche there," Dorothy confided to a friend. "I am venturing to ask you to direct me in this quest [for financial support], since I really don't yet know my way about in my new country. Also there is this ever present opposition of which you know—which must be circumvented if I am ever to have my chance to show what I can do."[8]

To Fan she wrote more frankly: "I hate S[mith], not to speak of A[mherst]...I want adult students and some nice colleagues. Neither here."[9]

As part of this broad program, Dorothy applied for a Guggenheim Fellowship to carry out "an inquiry into the mathematical foundations of a methodological approach to the interpretation of X-ray data obtained from molecular crystals." The end product would be a sequel to her 1946 monograph, working out the connections between this recent work on structure factors and the algorithm for deciphering point set Pattersons she had pioneered in 1939.

Character and Personality, Scholarship, Capacity for Independent Research, Ability to Make a Noteworthy Contribution, reads the Guggenheim checklist for those writing letters of support. Irving Langmuir, Martin Buerger, David Harker, and the chemist Kasimir Fajans all testified to Dorothy's sterling qualities, and the "extremely independent and strong character which has enabled her to continue work in her chosen field in spite of difficulties."[10]

This proposal, Langmuir explained, is one front in a two-front attack on the protein structure problem.[11] Dorothy would lay "a proper mathematical foundation by which Patterson diagrams can be calculated for types of atomic arrangements which probably exist in proteins." Computation was the other front: "to develop rapid calculating machines, perhaps eventually of the electronic type, which can speed up by a factor of a hundred or so the numerical calculations that will be necessary." The whole plan, he thought, should take about five years.

The Guggenheim selection committee met in New York early in 1947. Dorothy's proposal was rejected; the foundation won't say why. Then as now, the competition was fierce. Perhaps her age—52—weighed against her. Or was it her gender? Female Fellows were rare. Or was she blackballed out of hand—had "the incident" wafted to Guggenheim headquarters? One thing is certain. Linus Pauling was a member of the selection committee.[12] This could have been his moment of grace. He chose not to take it.

The Schenectady meeting von Neumann had proposed for January 1947 took place at last on April 16. Von Neumann, Langmuir, Dorothy, Harker, and Booth together pondered the feasibility of building metal protein molecule models, diffracting short waves with them, and comparing the diffraction patterns with the x-ray diffraction patterns produced by actual protein crystals. What Johnny wanted he didn't get after all. Reluctantly, the group concluded that even the computers he and Booth were designing would be no match for such heavy computations. Not then, not any time soon. Dorothy would not move to Princeton.

None of them could then imagine—who could have possibly imagined?—that by the end of the century computers would solve protein structures in a matter of hours and modeling and computer graphics would transform the mathematics-biology interface. Of the five, only Booth lived to see this.

Booth stayed with the Langmuirs for a few days after the meeting. "Very proudly, Langmuir showed me his 'electric pig.' Now called a garbage disposal. He claimed that this one was his own invention and that it would never jam. I had just eaten a banana and most unwisely threw the skin into the pig. It froze solid! I then had the interesting experience of crawling on the floor to help this Nobel Prize winner take the device apart."

Back at Birkbeck again, Booth invented the world's first magnetic computer storage device, now in the Science Museum in London. In 1953, he found an algorithm that speeded computations significantly. Then and now: the Booth Multiplier is used in Pentium 4 processors. The Booths emigrated to Canada in 1962, where they produced a new generation of computers, catalyzed research at several universities, and tackled the challenges of machine translation. When Booth died in 2009, they had worked together for 60 years.

Von Neumann didn't give up either. An avid cold warrior, he did not blink at the prospect of a hot one. The computer and the H-bomb were developed hand-in-hand: his.

In the 1940s, David Harker found a crystal structure-solving shortcut. He and his GE colleague John Kasper developed it further and used it to solve the structure of a poorly understood molecule, decaborane: "they stunned the crystallographic community…Their results ultimately forced a revision of Pauling's theory of chemical bonding."[13]

Harker and Kasper had discovered a new route around the phase problem: exploiting numerical relations among the intensity data. Today's crystallographer makes the measurements, feeds them to a computer, and goes out for coffee. In fact, you can leave earlier: the measuring and feeding are automated too. X-ray machines aren't cameras anymore, they're Geiger-like diffractometers. No more trial and error; no more Pattersons either. Today structures are solved by computerized "direct methods," which grew out of Harker and Kasper's work.

"The production of modern computers has strongly contributed to the rapidity and efficiency of their methods," said the Royal Swedish Academy of Sciences, awarding the 1985 Nobel Prize in Chemistry to Herbert Hauptman and Jerome Karle (a prize Booth felt Harker should have shared). "These methods are now so efficient that structure determinations for which the Nobel Prize was awarded in 1964 can today be made by a clever beginner." The 1964 Nobel had gone to Dorothy Crowfoot Hodgkin "for her determinations by X-ray techniques of the structures of important biochemical substances," most notably penicillin and vitamin B_{12}, crowning research she'd pursued for 30 years.

Before computers, crystallographers had ridiculed direct methods and their developers, in language now familiar. "Their method is no more useful than other direct methods already in use," snarled one.[14] "Largely fallacious," declared another. "I think," said Hauptman in a post-prize interview, "that the problem also was that most crystallographers did not have a strong background in mathematics, and many of them mistrusted mathematics. They simply could not believe what the mathematics was saying."[15]

In 1950, David Harker launched a Protein Structure Project at Brooklyn Polytechnic, funded by money from the Langmuir family and the Rockefeller Foundation. Seventeen years later, he solved the structure of the protein ribonuclease. John Kendrew had solved myoglobin and Max Perutz hemoglobin in the late 1950s (they shared the 1962 Nobel Prize in Chemistry). Thus Harker did not "win" the "race." In hindsight, that matters little. Harker's project developed important new tools and methods. One was the transition from film to diffractometers.

Old friends stopped off to see Dorothy on visits to New England: Dorothy Hodgkin, Margery Fry, Bertrand Russell, Katharine Blodgett. For several years after his Princeton stay, Booth returned to the United States in the summer, always with a few days in Woods Hole. One time that didn't work out. "Dorothy had arranged the usual collaborative effort," he told me. "She had obtained a superb set of diffraction values for a protein sample. This came from Fan. The project was to show that the spectra were in agreement with Dorothy's cyclol hypothesis. I undertook a comparison with several models and found that there was no way in which the data would fit a cyclol hypothesis but that there was reasonable agreement with a spiral arrangement. Dorothy did not like this at all so we did not publish the results. This undoubtedly lost us at least a part in the Nobel Prize."

In 1960, David Harker encountered Dorothy at a meeting in Cambridge, England, where John Kendrew and Max Perutz were showing their models of hemoglobin and myoglobin molecules. They looked nothing like cyclols. "And there was Dorothy Wrinch, chatting gaily with people, and the models were on a table about 10 feet behind her. I said, 'Oh, hello, Dorothy. So the structure is now really known, and there is the model. Do you want to look at it?' She said, 'No, I do not want to look at that model!' And she really didn't look at the model until much later. She was still hoping that it was wrong."

But Dorothy's bitterness eased somewhat. She made peace with Smith, if not with Amherst. After Otto's death she moved to a faculty suite in Hubbard House, one of Smith's homelike dormitories. With Office of Naval Research support, she wrote a stream of brief papers on Fourier transforms and on crystal twins, and taught a generation of Smith students. "She was very sweet to me as a student and very nice," said Elizabeth Moore. "I actually knew nothing about the cyclol business... I knew her work on crystals, twinning in structures and Fourier transforms, and the things that were really positive that she did—that she made contributions to."[16]

"Dr. Wrinch was the faculty resident at Hubbard House when I was there (1962–66)," wrote Nancy Reynolds Davidson. "She absolutely charmed my father on Fathers' Weekend freshman year. I introduced them in the dining room and, as they shook hands, she remarked, 'Surely this isn't the father. He's far too young.' We didn't have many, if any, science students in the house at the time. But we all knew that she was scary smart, devoted to her peptides and would probably win a Nobel Prize one day."[17]

As Faculty Resident, Dorothy presided over a dinner table. "October 1962," Nan Fitzpatrick recalled. "I can still see Dr. Wrinch sitting at the head of the round table in the Hubbard House dining room facing west, napkin boxes to the right. The Cuban Missile Crisis was in full swing. Our house president had spent the summer in Marine training at Quantico. She had Hubbard House in high alert and lined up and marching. That evening at dinner Dr. Wrinch dolloped out our entree, each with an equal part of the

Dorothy Wrinch with students in Smith College special studies class, "An Introduction to Molecular Biology," 1965–66. Left to right: Susan Bates, '66, Sylvia Fan, '66, Dorothy Wrinch, and Jane Rogers '66. Photographer unknown; copyright Smith College. The Dorothy Wrinch Papers, Smith College Archives.

parsley garnish. At dinner's end, as cool as a cucumber, she raised her parsley sprig and announced, 'Eat your parsley. It cleanses the palette.'"

But I'm told that some students avoided her table because she expected intelligent conversation.

"I took an interim course with Dorothy in my junior or senior year, which would have been January 1963 or 1964," said Barbara Zitzewitz. "I was there to learn some group theory but came away with more than that. She talked about her experiences during World War II and helped me see what a career in science would be like for a woman—how much dedication and self-sacrifice would be required. She and Gladys A (cannot remember her last name) were senior women adjunct faculty and the only women in science I encountered at Smith. There were no women faculty in chemistry when I was there. They were our role models."

"She had more of an impact on my choice of graduate studies than I realized at the time," Barbara continued. "After leaving Smith, I went to Harvard where I worked for 2.5 years in the doctoral program on the shapes of molecules. The work I started with Dorothy was continued there....I chose to go to Harvard so that I could work on the synthesis of organic molecules with simple geometric structures. E. J. Corey was engaged in that research and later won the Nobel Prize for his efforts. Unfortunately, Corey was not interested in having female students so I was unable to follow through with my plans. I left the chem department with an MA in chem and went to the ed school to complete an MAT in science ed."

Dorothy moved to the apartment on West Street in 1966. "The first Halloween we somewhat homesick freshwomen left Hubbard House and went trick or treating and had heard that Dr. Wrinch's home was a 'must,'" said Margi Hollingshead. "What a delightful woman! She invited us in and plied us with questions about our studies and our adjustment to life at Smith. We had heard through the grapevine that she had been Bertrand Russell's lover,[18] which made our visit with her all the more tantalizing and intriguing, some-how (in those conservative days, I'd never known anyone who'd really done that!). We stayed for quite a visit and left feeling included, cared for, and 'grown up'—special and proud to be Smithies."

STRANGE DOINGS AT SANDOZ

Alkaloids, the Nobel Foundation's website explains, are a numerous group of nitrogenous basic substances from the vegetable kingdom. Morphine, cocaine, and strychnine are among the more famous. The Nobel chemistry laureate in 1947, Dorothy's by-then enemy Robert Robinson, never thought to look for cyclols among the alkaloids, his own turf. Indeed, he ceded the ergot alkaloids to his friend Arthur Stoll, a chemist in the Sandoz Laboratories in Basel. Stoll found cyclols in ergots in 1951.[1]

Ergot, a fungus of barley and rye, is to hallucinations as morphine is to dreams. Its powers were known to ancient cultures. It is thought—Albert Hofmann, the discoverer of LSD, is one who thought—that ergot fueled the ecstatic rituals of the Eleusinian Mysteries, the ancient Greek festival honoring Demeter, the goddess of grains.[2] And in a compelling piece of scientific, literary, and linguistic detective work, historian Mott Greene identifies ergot with the *soma* of the ancient Sanskrit Vedas.[3]

The ancients knew, from received tradition and direct experience, the many intricate steps in making ergot safe for trips and treatments. For, ingested raw, ergot is deadly and the death is horrible. "Ergot contains a virulent poison called Ergotamine (or Ergotoxine)," Greene explains.

> The unfortunates who eat bread made from infected flour develop a disease called ergotism, also known as "St Anthony's Fire." It has two forms, convulsive and gangrenous. In convulsive ergotism, which may drag on for months, one experiences violent and terribly painful convulsions, blindness, deafness, diarrhea, and the opening of watery pustules on the extremities. Later, one dies. The gangrenous form begins with lumbar pain and limb pain and swelling, followed by violent burning pain (St. Anthony's

Fire), alternation of intense heat and cold, followed by gangrenous attrition and blackening of limbs, and emaciation. Then one's arms and legs fall off. Then one dies.

That's how 40,000 people died horribly in Aquitaine and Limoges in France in the year 994, and 12,000 in Cambrai in 1129. But no one blamed bread at the time; victims saw their suffering as divine punishment.

The ancients also knew ergot's virtues: in proper doses, it stimulates uterine contractions. The Chinese documented its use in obstetrics in 1100 B.C.E.; Methergine, a modern ergot drug that stops postnatal bleeding, was developed at Sandoz.

Ergot is fragile and extremely sensitive to chemical agents, to light, and to air. Moreover, as Leopold Růžička explained in his memoir of Stoll,[4] its pharmacologically active compounds are difficult to separate. Arthur Stoll was Sandoz's chief chemist and vice president when he solved the ergot structure. He'd been trying for 34 years.

Browsing in the library a few years after Stoll's discovery, Milton Soffer, the Smith chemistry professor, came across Stoll's paper.[5] "A linkage between amide groups of neighboring peptide chains, corresponding to that present in this formulation, had been previously suggested by Wrinch in connection with her cyclol theory of the structure of peptides," he read. "Subsequently, a number of objections were raised to this cyclol theory. Nevertheless, at least in this simple example of the peptide residue in ergot alkaloids, experimental confirmation of the theory of Wrinch seems to have been obtained; and as research enables us to penetrate further into the constitution of the peptides, it may well be that analogous structures will also be found in compounds of low molecular weight belonging to this class."

Though Stoll's paper was two years old, Soffer was the first to tell Dorothy. Perhaps her circle didn't overlap the *Progress in the Chemistry of Organic Natural Products IX*'s readership. Or perhaps the cyclol controversy was history, forgotten, and Dorothy a small-font footnote. Or perhaps she had antagonized them, every one. "Why Eric," she complained to Neville, "in God's name did you never tell me that Robinson had been talking about 'DW shoving her old cyclols down your throat'?"

Letters she wrote after Otto's death display her worst qualities: outspoken, and spoken out, contempt for the mistakes of others; still-burning and bitter resentment of the slightest of ancient slights. They are painful to read. Why, she demanded of Harker, had Patterson been so rude to her at a recent meeting? Because, he replied bluntly, you had driven him over the edge, arguing every point, not letting anyone get a word in, not even letting them give their own talks. I have defended you all these years though everyone says no one as biased as you can have anything useful to say, but now even I'm fed up. Of course your criticisms are useful in their way, but why must you play such a negative role?

But now Stoll had found her cyclol rings. Not cyclol fabrics, not cyclol cages; these have never been found. But the rings had been the epicenter of the fray. Pauling had "proved" they couldn't exist, yet here they were, at the unassailable Sandoz, brute experimental facts, plain as day. And, said Stoll, there may be cyclols in other peptides.

Arthur Stoll, born in a Swiss village in 1887, studied chemistry in Zürich and moved to Berlin with his teacher, the famous chemist Richard Willstätter. Stoll shared in the work on photosynthesis for which Willstätter won the Nobel Prize in 1915, the year the Braggs won the prize in physics. He did not share in the prize itself, but the Sandoz company noticed him and lured him to Basel to create and direct their pharmaceutical department.

Stoll isolated ergotamine, a pure alkaloid, right away. "Five further alkaloids, most of them to Stoll's credit, could be isolated from ergot," wrote Růžička. But "it proved considerably harder to determine the structure of the ergot alkaloids themselves."

They have two components, Stoll showed: lysergic acid and a peptide portion. "Chemists investigating their structure had three main questions to settle," he explained, "the structure of the lysergic acid portion of the molecule, the structure of the peptide portion [where the cyclols are], and the nature of the linkage between the two portions."

For Sandoz, the ergot structure was a means to an end: drug production. The company had grown rapidly from humble beginnings in 1886. The small dye manufacturer first added a fever-reducing drug, then the sugar substitute saccharine. It mushroomed between the World Wars, producing more drugs and chemicals for textiles, paper, leather, and agriculture.[6]

In the late 1930s, while Dorothy and Dorothy battled over insulin, Albert Hofmann began searching for a respiratory and circulatory stimulant. He synthesized lysergic acid and created 25 compounds, Methergine among them. The last of his series, LSD-25, did not produce the effects he was looking for. Had he followed Sandoz procedures, he would have consigned it to the research wastebasket. Some sixth sense told him not to.

Hofmann would tell, and retell, the story to the end of his very long life. One April afternoon in 1943, "being affected by a remarkable restlessness, combined with a slight dizziness," he went home and lay down. There he "perceived an uninterrupted stream of fantastic pictures, extraordinary shapes with intense kaleidoscopic play of colors. After some two hours this condition faded away."[7] He had not intentionally ingested LSD-25 or anything else. Maybe he had accidentally absorbed a minute amount through his skin, he told the company pharmacologist. The pharmacologist tried it himself; the effect was reproduced faithfully.[8] Stoll and Hofmann received U.S. Patent 2,438,259 for LSD on Mar. 23, 1948. (The suffix 25 was dropped.) Sandoz became the world's largest—indeed for a time the world's only—manufacturer of LSD.

Protein chemists dismissed the ergots, but other chemists did not. "Today the beautiful atomic patterns of these alkaloids have been established," Dorothy wrote in the *New York Herald Tribune*,[9] "and synthetic studies are reaching a final stage, resulting from extensive researches by Stoll and his co-workers, and by Jacobs and Craig, Woodward and his co-workers in America, and many others. The results include products of wide practical significance in internal medicine and neurology; new drugs have been synthesized for use in the treatment of hypertension, peripheral vascular disorders, angina pectoris and other disorders." Cyclol chemistry became a respected if minor field of research. Papers poured forth, some from the Robert Robinson Laboratories in Liverpool. Dorothy flew back and forth to England.

In England she visited her sister Muriel, who had followed her now-grown daughters back to her homeland. Talitha and her husband Nichol Williams were farmers in Wales; Carla, a nurse and a writer, and her husband Valentine Blake ran the Duke of Devonshire's fishing estate in Galway, Ireland. Rosalie, the sculptor, lived near London with her husband, Johnny Johnson, a Fleet Street journalist. Muriel lived in a cottage her daughters bought for her and ran a bed-and-breakfast inn. But she missed the South African sun and her two sons, who had stayed behind.

From London, Dorothy wrote gaily to Fan:[10]

Am having a lovely time (except for the WEATHER) and have been made an hon. res. associate in biochemical department at University College London which is one of my old loves since my very first job was there (in maths dept). It is SO nice to be invited into the fold and given a party of welcome, etc. IMAGINE!!! I was so pleased.

and

Preparations are being made for Bertie Russell's 90th birthday which is to be celebrated in Festival Hall on the right evening (May 18). AM hoping to go since he was once my greatest hero. Have some new heroes since I arrived—organic-cum-pharmacological chemists—so nice.

Russell was the Man of the Century, she later told his archivist, and the greatest beneficial influence on her personal and scientific life.

One new hero was M. M. Shemyakin in Moscow, for whom the Institute of Bioorganic Chemistry of the Russian Academy of Sciences is named. Cyclol chemistry was one of its specialties.

Dorothy spoke at conferences and kept up with the burgeoning cyclol literature, but contributed little that was new. She saw her legacy in grander terms.

Under the table, Sandoz and the CIA worked hand-in-glove. In a roundabout way, their strange doings sprang me onto my path to Dorothy.

The CIA saw possibilities for mind control, but LSD wasn't only for spooks. LSD induced schizophrenia-like behavior; this "model psychosis" might hold clues to the disease. One of the early LSD-as-schizophrenia researchers was Arthur Stoll's son Werner, a psychiatrist at the University of Zürich. My father, at the Addiction Research Center in faraway Kentucky, was another.

Robinson's alkaloids—morphine, cocaine—were Dad's bread and butter. He and his colleagues also tested, for their paymaster the U.S. government, drugs like codeine, Demerol, amphetamines, tranquilizers, and barbiturates to determine whether they were addictive, and to establish medical, nonaddictive doses. Methadone was on that list, provoking the flap with Marie. The CIA added LSD, and Sandoz sent it to Kentucky.

Dad found that LSD does not mimic schizophrenia: after a few days of chronic intoxication, its effects disappear. But, he noted, LSD does mimic mescaline.[11]

I was a restive teenager at the time. Basketball was King and I was sportsphobic. Nor could I stomach the hypocrisy I saw around me in Lexington: locking people up for possessing alkaloids while solid citizens grew rich raising tobacco and distilling whiskey and then blowing it at the races. My solace was a weekly piano lesson with a gifted pianist and composer. Helen Lipscomb was wheelchair-bound, a victim of polio. She lived with her mother; her father had died. Her brother Bill—the same William Lipscomb who had reviewed Dorothy's monograph and would win the Nobel Prize—was then an up-and-coming chemist and crystallographer in Minnesota. We, Helen's adoring students, drew a music circle around her and gave recitals in her living room.

Music notwithstanding, in my junior year I resolved to get out: out of the house, out of Lafayette High, out of Lexington, out of Kentucky. The sympathetic Dean of Girls told me that the University of Chicago took bright kids bored with high school. I did not know, nor would I have cared then, that UC had one of the world's greatest math departments. My one ambition was to escape a life of afternoon bridge.

You'll miss the prom! my astonished friends wailed. Exactly.

My parents growled; we fought for weeks. I won, suddenly and unexpectedly, when Dad remembered that Heinrich Klüver was on the UC faculty. The famous Heinrich Klüver, the world authority on the hallucinogen mescaline! A college with such an outstanding scientist in its ranks might be worth the steep tuition, $230 a quarter.

Dorothy had a new mission now, a higher mission: weaving a logical map of the atomic architecture of life. Every fact would take its place in this map, and every set of facts would find its proper class. The spirit of her pseudonym, Jean Ayling, that voracious synthesizer of preprints, reprints, and reports, repossessed her in full force. No matter that her grand scheme for Child Rearing Services had never been realized, that her own experience with child rearing had been stressful and fraught with ambivalence.

"This is it," she wrote, "an account which satisfies me of how the protein problem can be pictured and how with this picture we can see the integration into a single whole of many different issues and theses, in physiology, normal and abnormal, in morphology, in pharmacology, cytology, enzymology, immunology, high polymer chemistry and organic synthetic chemistry…crystallography, etc., etc."[12]

She would weave her own strands together too: her discipleship in logic, her expertise in mathematical physics, her theory of scientific method (intact despite decades of struggles with actual scientists), her immersion in chemistry and biology, her mastery of diffraction geometry, her studies of crystal symmetry, her lifelong search for patterns. *Chemical Aspects of the Structures of Small Peptides* would be her *apologia pro vita sua.*

Eric Neville, wracked by illness, found a publisher and saw it through the press.[13] Dorothy proudly presented a copy to her niece Rosalie. "Breakfast reading," says Rosalie.

Chemists couldn't read it either. That Dorothy still spoke their language with a foreign accent was surely part of their difficulty. She had anticipated that. "There may well be short passages, etc., and various statements which may be capable of misinterpretation and may in fact alienate people at first glance," she wrote to Harry Sobotka.[14] "I naturally have to be ready to cut out…anything…you may adjudge should be cut out rather than give people the chance to say, 'you have only to look at pp.…to see that she just doesn't *know* things that any chemist knows."

Couldn't read it, or wouldn't? "Suspicion of speculative chemistry ('paper chemistry') continues to be shared by most of the leading chemical journals, which refuse up to this day contributions containing no new experimental results," Michael Polanyi had warned. "In spite of the fact that chemistry is largely based on the speculations by Dalton, Kekulé and van't Hoff, which were initially unaccompanied by any experimental observations, chemists still remain suspicious of this kind of work."[15] As if on cue, the reviewer for the *Journal of the American Chemical Society* complained, "No original experimental work, chemical or otherwise is reported. What might have been presented simply and concisely in a dozen pages has been expanded to almost two hundred by the use of a maze of symbolism and grandiloquence."[16]

One reviewer, an organic chemist, put his finger on the problem: Dorothy wrote for the Aristotelians. "It is most unlikely that the views now advanced by Dr. Wrinch will command widespread acceptance," he said.

> The reason for this is fundamentally semantic. Dr. Wrinch starts from the undoubted fact that in certain compounds containing amino-acid residues…these residues are not combined through simple amide linkages.…She then argues that the existence of such compounds "means the repudiation of the peptide hypothesis, for if a hypothesis is subject to unpredictable exceptions, no predictions from it are reliable," and proceeds to a detailed examination of all the possible ways in which amino-acid residues might be combined, arriving finally at a generalized structural theory in which the "cyclol" and "amide" structures figure as special cases. Such an approach is however, mathematical rather than chemical, and will not be accepted by organic chemists.[17]

Dorothy referred to this book as "Mon One" because she was planning a sequel. There she would unfurl the map in full glory, a Grammar of Life. "Mon One is, in my scale of reckoning, a digit against something in the hundreds of thousands for Prot Mon," she told Neville.

But for the first time in her life she was daunted. "I have been looking forward to the time which is now here and when the road is clear regarding time, money, etc. to get the Prot Mon done—and now I can't do it. Isn't it hell.... Why can't I get on with it?"

Well, all I have to say, my dear little Dorothy, is that you will come to a bad end. Half a century later the headmistress's voice still burned her ears. "Turn my memory back and try to analyse what she meant (if, in fact, she ever did say this)," Dorothy mused at the age of 70. "She was, as I remember, talking about being public-spirited—school motto was 'What touches one, touches all'—about being 'selfless in serving the school,' in putting the welfare of the school above my desire (for example) to learn mathematics. I judge now, from this long distance away, that her statement was in terms of 'playing the game,' 'supporting your school, your church' etc. etc."

Dorothy consulted a therapist and kept a notebook.[18] Pam was struggling to write her Ph.D. thesis in international relations at Yale.

The same with Pam I suppose...and me. The alternatives are not...being a protein Newton and being mud. There are lots of possibilities between.

They are not...writing the longest biggest...thesis ever written or nothing.

NB Dan says: To have very high evaluations of things in a highly competitive society is a sure way to misery...(Hitch your wagon to a star—any child can be president...) This excessively high evaluations inculcated...church state home and school...all the time.

Chemical Aspects of Polypeptide Chain Structures and the Cyclol Theory was published at last in 1965.[19] Neville did not live to see it.

"If a reviewer had been asked to consider a book on the 'phlogiston' theory years after Lavoisier had shown that it was incorrect, he might have felt the same way as I have with this one," says the one review I've found. "It would be unfair not to admit that there may be important and useful hints for future work in this compendium, but they were not evident to this reviewer."[20]

Dorothy's copy of *Grammar of Ornament* passed to me. Egyptian tomb ceilings, wall tiles in Grenada, Indian rugs, tattoos of savage tribes: Man the Pattern-Maker has ornamented every surface, in every era, in every place, since we first taught ourselves to carve and paint. Now I wonder why.

Owen Jones, the author of *Grammar of Ornament*, likened pattern making to worship. The astronomer Sir John Herschel, Jones's contemporary, thought our drive to make patterns might be innate. Are we born with "a kaleidoscopic power in the sensorium to form regular patterns by the symmetrical combination of casual elements?" he asked.[21] Is symmetry all in our heads?

Reading everything I could find on the origins of ornamental patterns, I came across a little book, *Mescal,* written in the 1920s. Its author was Heinrich Klüver.[22] The name was somehow familiar; I opened it with a puzzling sense, not of déjà vu—I had never seen the book before—but *déjà rencontré.* Then it all came back to me: this man had changed my life.

In my four years at the University of Chicago, I never met the famous professor, my unwitting *sine qua non.* No one did. He worked alone in his basement laboratory and never taught a course, graduate or under. What was he doing down there? In *Mescal* he had described images seen under the influence. He'd seen them himself; he ingested mescaline in carefully measured doses and recorded every sensation, every flicker. Four types of geometric forms appeared sequentially and predictably: first gratings, lattices, fretworks, filigrees, honeycombs, or chessboards; then cobwebs; then tunnels, funnels, alleys, cones, or vessels; and last, spirals. "Many phenomena are on close examination, nothing but modifications and transformations of these basic forms," he wrote. "The forms are frequently repeated, combined, or elaborated into ornamental designs and mosaics of various kinds."

Moreover, these designs and mosaics appear, to the hallucinating mind, to be projected on a ceiling! This might bear some relation to the origins of ornamental art, Klüver suggested.

In *The Mind in the Cave,* a study of the prehistoric cave art of Spain and France, David Lewis-Williams argues that the artists were hallucinating shamans. He cites Klüver and updates the neuroscience. "Simply put," he says, "there is a spatial relationship between the retina and the visual cortex: points that are close together on the retina lead to the firing of comparably placed neurons in the cortex. When this process is reversed, as following the ingestion of psychotropic substances, the pattern in the cortex is perceived as a visual percept. In other words, people in this condition are seeing the structure of their own brains."[23]

Forty years earlier, in my first lecture on symmetry, I thought I was speaking to everyone but the psychologists. Now I realized it was only for them.

In Paris not long ago, I revisited the Notre Dame Cathedral. Again I was awed by its magnificent stained-glass windows, the enormous multicolored rosettes, but where once I had seen a medieval kaleidoscope, the ultimate epitome of abstract symmetry, now each pane and panel has meaning, each tells a different story. I can no longer divorce the impression it makes from the craftsman's skill.

But neither can I divorce the craftsman's skill from the impression it makes. How else, but through pattern, shape and form, can I grasp meaning and story? My thinking has come full circle. Beauty is truth, truth beauty, said Keats. Emily Dickinson concurred: *we two are one*. But I see beauty and truth as warring twins joined at the hip, heart, and head. Symmetry must be explained—but if not for its beauty, why would we try to explain it?

And so as kinsmen met a night,
We talked between the rooms.

23 SWAN SONG

Pamela Nicholson Wrinch (her names legally permuted) finished her thesis, wrote a monograph titled *The Military Strategy of Winston Churchill*,[1] and then taught in several colleges and universities in the Boston area. Dorothy implored her to get married. Approaching 40, she did. Alfred Schenkman, a Cambridge (Massachusetts) publisher, had a sturdy build, wide smile, and dark hair. He was smart and charming and casual with money; Dorothy paid for the roast beef for their wedding reception. The Schenkman Publishing Company, then five years old, featured sociology and foreign affairs; it published women's studies and Black studies before bookstores knew where to put them. Dorothy sent a box of Schenkman books to her old friend Dorothy Needham.

I spoke with Pamela on the phone a few times, but I never met her. Once I asked Dorothy if her daughter resembled her. "Oh, no, poor dear," she replied, "she looks like…" and her voice trailed off. In family photographs, Pamela does resemble her father, as he looked in 1913, shy, dark, and handsome.

Dorothy, when I first met her, sparkled with wit and energy. She swam, played the piano—four hands when she found a partner—and visited Pamela and Alfred in Cambridge on weekends. But gradually she flew to London less often, and summers in Woods Hole lasted further into the fall. Gladys Anslow, her dearest friend in the Smith faculty, died of cancer in 1969. Her sister Muriel had died a few years before. A pilgrim through the world's religions, she recounted her long spiritual quest in a book called *Key to Living*. She wanted, above all, to visit Greece. Her son Peter gave her steamer tickets for her birthday. She died at sea and was buried in Corfu.

In 1972, the Smith trustees voted to give Dorothy an honorary degree. She was living year-round in Woods Hole by then; President Mendenhall asked Pamela to bring her to Smith for the ceremony. Pamela agreed, but Dorothy couldn't or wouldn't come, and honorary degrees aren't given in absentia.

I visited Dorothy in Woods Hole a few times. We talked in her sunny living room. She was losing interest in our book and in crystals, I could see. What had these elegant forms, their visible and hidden symmetries, meant to her? Herman Weyl, Einstein's colleague at the Institute for Advanced Study, closed his career with a series of lectures on symmetry.[2] Of his own swan song he wrote:

> I felt like a man who labored through a long day's work, doing his share as well as he could in the conflict of ideas and human demands, and who now, as the sun is sinking and the conciliating night approaches, plays himself a quiet evening song on his flute.[3]

Dorothy had begun writing the book on crystals before I knocked at her door. Luck had sent me, and the luck was mine: if you want to learn a new field, don't start with bland textbooks. Find an old hand with a viewpoint.

Dorothy had a viewpoint, the viewpoint of the famous Swiss crystallographer Paul Niggli. In a note she sent Warren Weaver from Zurich in 1937, Niggli got more lines than Růžička.[4] Niggli had "very kindly applied his great knowledge of crystal forms to my cyclol molecules and suggested many relevant facts, etc., which fit in very well," she wrote. "He really does agree with me in thinking that practically *all* the strange things which characterise the cyclol structures have exact precedents in the inorganic world—even the cage-like form of the molecules."

What insights, she asked, can we glean from unusual crystal forms? Protein crystals, like many minerals, often grow in pairs (or triplets). Some are conjoined like Siamese twins. Others form a cross or look like cubes grown through each other.

Mineralogists call such groupings—whatever the number of individuals—crystal twins. To distinguish twins from haphazardly intergrown crystals, they laid down criteria long ago. First, to count as a twin the configuration must occur often, often enough to rule out chance. Next, the configuration must have a clear geometric description, a "twin law," a symmetry operation (reflection in a mirror, rotation around an axis) that relates the individuals to one another. A third criterion is a post-1912 postscript re-imagining the second on the invisible atomic scale: a shared row of atoms, or a shared lattice plane, or a shared three-dimensional array.

Dorothy didn't challenge these criteria, she seized them. The shared lattice elements are "themes," she said. They are the keys to protein structure. She wrote a few papers on mineral twins in the 1940s.[5] When she learned that Stoll had found her cyclols in the ergots, she set twins aside. But now, her two Mons published, she returned to them.

Those interlocked cubes and crosses obsessed me from the day I met her. I examined mineral specimens, drew projections, built models, and stared at old drawings in great yellowing tomes like *Der Diamant*. I scoured mineralogical and crystallographic journal articles for theoretical insights. I found two main theories. They didn't agree. The Beauty school of thought viewed twinning through the lens of symmetry and form, the perfect peace of the laboratory. The lens for the Truth school was real crystal growth, a rumble of defects, inclusions, dislocations.[6] D'Arcy Thompson had assumed that growth and form shed light on one other. But the more I read, the more it seemed that each made the other more mysterious. I wished I could talk with Niggli, the great mineral morphologist. But he had died in 1953, so I went to Groningen instead. Home again, I wrote a critique of the twinning literature.[7] Growth comes first, I said. And crystal growth, like politics, is local. Dorothy would have hated it.

A few years later, at the crystallography institute in Moscow, I lost what little faith in symmetry I had brought with me. I no longer believed my words in the Symmetry Festival guidebook: *in science as well as in the arts, symmetry is the geometric plan on which the variations of nature and of life are drawn.* I still saw beauty in symmetry, but no inexorability, and no significance. In a lecture to my Smith colleagues when I returned, I proposed a Festival of Disorder.[8]

In the fall of 1975, Pamela moved Dorothy from Woods Hole to a nursing home near Cambridge. The place was large and cheerful and the lounges were comfortably furnished, but I found it painfully depressing. Elderly residents, some in wheelchairs, others pushing walkers, milled vaguely about, staring at television sets or at visions of their own. Dorothy sat lost in a large armchair, oblivious to everyone and everything around her. She seemed to have shrunk. I touched her shoulder and she looked up. Did she recognize me? I wasn't sure. I talked with her, or rather to her, for a little while, then drove home.

I called Pamela the next day. Did Dorothy *have* to be in a place like that? In her own home, surrounded by her books, her little garden, she might take an interest in life again. Soon she was back in Woods Hole, with live-in care around the clock. Pamela visited on weekends; she had a summer house nearby.

One cold November weekend, Pamela's summer house caught fire. She was in it and did not survive. Alfred told me the cause was a faulty space heater, but I've also heard she was smoking in bed. The Falmouth Police Department did not respond to my inquiries.

Dorothy had to be told. I visited her a few weeks later. She lay on a sofa in her living room, her face toward the wall. She hadn't spoken since the fire, the housekeeper said.

Alfred carried the urn with Dorothy's ashes into the church and set it on the altar. A church! A church funeral for the Heretic, Russell's student, Og's pal, Dora's best friend? A church funeral for the eternal scoffer? Never mind. *Lord Dismiss Us with Thy Blessing* for the very last time. The click-click of wooden heels on wooden floors fades away.

A few years after the Symmetry Festival, Alfred offered me a deal: if I would organize a symposium to celebrate Dorothy's gift of her papers to Smith, he would publish the proceedings. Though he never sent Girton College the money Dorothy willed to it, he eventually kept his end of this bargain.

I invited Joseph Needham, but he refused to come. The last time he tried to visit the United States he was denied a visa, he said, and he would not ask again. Dorothy Hodgkin said she was too busy, but we exchanged a few letters. Why were the cyclol model and its creator *so* controversial? "Probably some of the intensity of feeling in the case of Dorothy Wrinch was due to the major importance of the scientific problem itself, the structure of proteins—nobody could be indifferent to the search for the truth about proteins," she wrote. Recently, among Hodgkin's papers in Oxford's Bodleian Library, I came across her handwritten first draft of that letter to me. The last phrase was written above one she'd crossed out: "and the success which she appeared to achieve was so outstanding." Delta as Icarus, wings melting in the fierce glare of jealousy.

Another source of trouble, Dorothy Hodgkin noted wisely, might be that the model had a catchy name.

David Harker, Martin Buerger, Carolyn Cohen, and Arthur Loeb (a crystallographer and Pamela and Alfred's neighbor) came. So did Woods Hole neighbors Ruth Hubbard and George Wald, the Dutch crystallographer Caroline MacGillavry, and other friends and former students. "I haven't had so much fun since the pigs ate grandma," Harker chuckled.

"More than most of our colleagues, she was eager to listen to us and to scrutinize the essential features of our work," I told the audience. "No unexamined platitudes were tolerated, no poorly drawn conclusions were allowed to stand. She demanded that we look at what was before us, that we recast our ideas in simple terms, and that we recognize their consequences. Dorothy never confused kindness with empty praise."

Reading my notes now, I see I was also describing my father.

Invited guests, Smith colleagues, students, drop-ins: we talked about Dorothy's broad interests and her wicked sense of humor. We talked about the compelling beauty of the cyclol model and whether aesthetics drives science. We talked about symmetry, *The Grammar of Ornament*, and patterns in music and dance. We talked about women in science, about Rosalind Franklin and *The Double Helix*, and the court of scientific opinion with its biases. She who kicks the hornets' nest gets stung.

We talked about obsession in science. Dorothy was not alone. Linus Pauling never stopped insisting on vitamin C's near-miraculous medical powers. Harold Jeffreys, one of the great geophysicists of the twentieth century, never accepted continental drift. The structure Robert Robinson proposed for penicillin was proved incorrect, yet he hoped, to the end of his life, that it was right after all. "I think she was right in thinking that the cyclol theory was going to be the great theory," said Buerger. "If this was correct, it would be the thing that we're feeding on now, but it turned out not to be the case." But, said Harker, "she fought unwisely…she would never give an inch, she would never alter her tack."

Harker pronounced the benediction:

> Her value to me, to all of us, is that her hypothesis produced so much interest in the structure of protein that now, about 40 years later, there are close to 100 structures known…it was a great attempt, produced tremendous scientific activity, and I think we owe a great deal of gratitude to anybody who can start the acquiring of such a large body of knowledge.

Today, 35 years after her death, the Worldwide Protein Data Bank holds detailed structure data for over 78,000 proteins.

The proceedings of the symposium, published as *Structures of Matter and Patterns in Science*, was not a best seller and is long out of print.[9] Rereading it now, I see we hardly knew her.

We'd read her papers; we thought she'd saved every scrap. We didn't know the stories in other archives.

The Dorothy we thought we knew seemed bewitched by symmetry. But symmetry was a stand-in for her magnificent obsession, the banner she carried through the cultures of science she encountered on her journey.

"Sire," said Marco Polo to the Emperor Kublai Khan, "now I have told you about all the cities I know."

"There is still one of which you never speak." Marco Polo bowed his head.
"Venice," the Khan said.

Marco smiled. "What else do you believe I have been talking to you about?"[10]

From the summation of pleasures to electrostatic mitosis to Mon One and Prot Mon, Dorothy never lost sight of the summit: when biology grows up it will be geometry.

But maybe the sciences are evolving in the opposite direction, away from geometry, dissolving into complexity, taut networks of laws and theories unmasked as human artifacts. Today the debate between John Scott Haldane and D'Arcy Thompson that young Dot witnessed in 1918 reads like *plus ça change* all over again. In the wise words of three generations of a family of physicists, "What seemed great problems to them [great scientists of the past] may seem trivial to us because we now have generations of experience to guide us; or, more likely, we have hidden them by cloaking them with words."[11]

"Causation and explanation don't always run upwards from lower to higher levels," says Denis Noble, professor of cardiovascular physiology at Oxford. "Indeed we may need to know about the higher levels in order to explain the lower-level data that form an input to the mechanisms involved."[12]

In her obituary in the London *Times*, Dorothy Hodgkin wrote, "I like to think of her as she was when I first knew her, gay and adventurous, courageous in the face of much misfortune, and very kind."

"Adventurous" was an afterthought. In her first draft, she'd written "obstinate."

24 EPILOGUE

David Harker and his wife Deborah returned to Northampton several times. On their last visit they brought me a cut-glass polyhedron.

The shape was the gift: a regular tetrahedron with its corners truncated, an ethereal rendering of Dorothy's cyclol C_1.

The polyhedron is also a three-dimensional kaleidoscope. Each triangular facet, finished in matte, reflects back and forth in the crystal-clear hexagonal facets to produce the illusion of an icosahedron. Amazing!

The icosahedron you think you see in this glass polyhedron is a ghost, the reflection of its triangular facets in the hexagonal ones. Polyhedron by Steuben Glass, gift of David and Deborah Harker. Photograph by Stan Sherer.

But…look closely. The icosahedron you see is as wretched as the paper hexahedron I made for Dorothy all those years ago. The edges are misaligned, the corners don't match up. This, however, is not the fault of the Steuben designers. There's no way around it: the tetrahedron and icosahedron live in different mathematical worlds. Rational and irrational, polyhedra that can be inscribed in lattices and polyhedra that can't be, crystallographic forms and non.

Change came suddenly, as Owen Jones had said it would, by throwing off some fixed trammel. That trammel was the crystal lattice.

I first heard of the ingenious Alan Mackay in Moscow. Alan is a crystallographer at Birkbeck College, London, now retired. A protégé of J. D. Bernal, he has the same fertile mind, right-on insights, and panoramic talents. Soviet crystallographers admired Alan for many reasons, among them his message: *Think beyond the lattice! To hell with the unit cell!* Alan didn't just think beyond it; he found an escape route. In 1976, he sent his friend Charles Taylor, a physicist at University College, Cardiff, a pattern to diffract. Taylor was an expert on optical diffraction and its x-ray analogues. With Henry Lipson, he had written a student-friendly textbook, *Fourier Transforms and X-ray Diffraction*. (Nothing new here, they wrote; "Wrinch's well-known book—*Fourier Transforms and Structure Factors*—was published in 1946. But Wrinch's book, we think, is rather too formal to appeal to non-mathematicians.")

Taylor knew—everyone knew—that sharp bright spots are telltale signs of an orderly array. The pattern Alan sent him was all pentagons, little ones, big ones, pentagons within pentagons. Profoundly noncrystallographic. Its diffraction photograph would, of course, be a blur.

But, said Alan, "I have tried the negatives on our diffractometer and they have given patterns which look very crystalline."

The pentagon-packed pattern was a Penrose tiling, discovered by the Oxford theoretical physicist Roger Penrose in the early 1970s. Penrose tiles are two quadrilaterals with rules for putting them together. Copies of these tiles can cover the infinite plane, but the pattern is odd. It appears to repeat, but never quite does. (See figure 11a.) Penrose proved that his tilings have no lattices. There's nothing like them in *The Grammar of Ornament*. But Owen Jones would agree: it's time for a new edition of that classic tome.

Penrose kept relatively quiet about his remarkable discovery until his tiles were patented. Then *Scientific American* featured them on its January 1977 cover, and Martin Gardner explained how they work in his column "Mathematical Games."

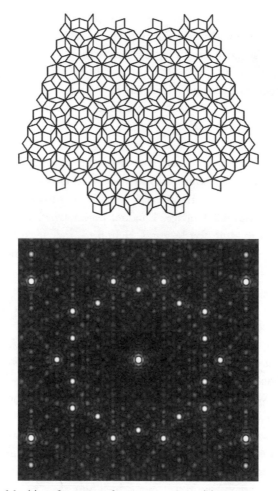

Figure 11 a and b (a) A fragment of a Penrose tiling. (b) A computer simulation of the diffraction pattern produced by this fragment. From M. Senechal, *Quasicrystals and Geometry*, Cambridge University Press, 1995.

Taylor did the optical diffraction as Alan requested. To his surprise, though the tiling has no lattice, the diffraction pattern had spots, sharp and clear (figure 11b). Sharp and clear—bespeaking *some* kind of order—order with "forbidden" symmetry! The message was as clear as the spots: order and lattice are not one and the same. Alan unveiled the photograph to a large, bemused audience at an international crystallography congress in Ottawa in 1981. I was there. None of us saw what was coming.

"Future progress will blend tradition with a fresh look at nature," said Owen Jones, and he was right on that score too. In 1982, Dan Shechtman, an

Israeli materials scientist working at the National Bureau of Standards,* created high-tech alloys of aluminum and manganese with diffraction patterns very much like Alan Mackay's.

Impossible, said Shechtman, I must have made a mistake. He tried again. The next batch had decagonal symmetry too; again impossible. Again and again, for two years; no one believed him until a senior colleague, John Cahn, realized Shechtman might be onto something. "Well, Danny, have you read *The Structure of Scientific Revolutions?*" Danny hadn't, but he did. And then he knew: he had smashed a stubborn paradigm, a foundation of the theory of the solid state. Late in 1984 Shechtman, Cahn and two colleagues announced the discovery in *Physical Review Letters*.[1]

Again I was lucky to be in the right place at the right time: with Louis Michel, a mathematical physicist and permanent member of the Institut des Hautes Etudes Scientifiques near Paris, I had organized an international conference on mathematical crystallography for the month of January 1985. Louis showed me the paper when I walked in the door. Coincidentally, Shechtman, Cahn, and co-author Denis Gratias were in Paris that month too; they hopped the RER-B to Bures-sur-Yvette. With them and our invited speakers and others who'd got wind of the discovery, we spent the month exploring Penrose tilings and the strange new alloys, and trying to understand how, if at all, they might be related. It was an exciting time. Everything we'd thought we knew was suddenly wrong. The "crystallographic restriction" wasn't a law of nature, it was just a theorem, a theorem about geometry! Suddenly we noticed that no one had ever proved, statistical mechanically, that lattices are the low-energy crystalline state. Lattices were a useful fiction after all. Now their usefulness was in question once again. Nature, even in the mineral kingdom, is more inventive than anyone supposed.

Quasicrystals, as Shechtman's alloys were called in deference to the defunct lattice paradigm, rocked solid-state science like an earthquake. They were replicated in laboratories around the world, and other forbidden symmetries were found in other alloys. *Shechtman had discovered a new state of matter!*

The months became years, the years decades. Scientists learned how to grow big quasicrystals, stable at room temperature. Behold! Some are perfect—perfectly regular—icosahedra and dodecahedra! You can see them yourself: images abound on the Internet.

A new field of mathematics, "long-range aperiodic order," was born and its problems were hard. Is order-disorder a dichotomy, or is it a spectrum?

---

*The National Bureau of Standards is now the NIST, the National Institute of Science and Technology.

Which arrangements of atoms produce which sorts of diffraction patterns—*can we tell the dancer from the dance?*[2]

The International Union of Crystallography created a commission to redefine "crystal." We—I was a member—argued. Some wanted to broaden the definition of lattice. I, and others, wanted to scrap the lattice altogether. We scrapped it: a crystal is any solid with sharp bright spots in its diffraction pattern, we said. Sharp bright spots, that's the criterion; if it looks like a duck and quacks like a duck, then it's a duck.

Very well, but what is a duck? Alan Mackay had raised that question implicitly. We, mathematicians and physicists and crystallographers, are still trying to answer it.

Icosahedral crystals? Nonsense! roared the king. Linus Pauling was nearing 90, but time had not mellowed or slowed him down, and two Nobel Prizes had not humbled him. Battle was still his cup of tea, his battery, his catalyst. After writing *Vitamin C and the Common Cold,* he'd forged on to promote ascorbic acid as a cure for cancer. Like Dorothy with her cyclols, Linus dismissed, with a wave of the hand, the mounds of studies that seemed to disprove him. Now he embarked on his last crusade: debunking quasicrystals.

Years before, after a lecture at Smith, Pauling had walked away from my question: if the atoms in a crystal aren't packed like a lattice, how else might they be packed? Now he took it on. Curiously he, who had once dismissed symmetry as subsidiary, argued for its primacy. A new state of matter? Hardly! These alloys are crystal twins, he insisted.

"There are tens of thousands of chemists in the United States, and Pauling was their star," Shechtman recalled.[3] "He would open the conferences of the American Chemical Society, and quasiperiodic crystals were always his topic. I attended one of the conferences, at Stanford. Thousands of people were there, and he attacked me. He would stand on those platforms and declare, 'Danny Shechtman is talking nonsense. There is no such thing as quasicrystals, only quasi-scientists.'"

In March 2011, the Technion celebrated Shechtman's 70th birthday with an international conference on the new scientific world disorder. I told him about Dorothy Wrinch and asked him how Pauling had affected his career. "I thought he was destroying it," he replied. "I was unknown and he was powerful, and here he was tearing me apart. But then I realized, it's just the opposite! This is *good* for my career! The greatest chemist in the world is attacking *me*! It shows this is important! If he hadn't done that, who would have paid attention?"

Pauling died in 1994, disbelieving to the end, but today even his staunchest admirers admit he was wrong about quasicrystals. Shechtman's office wall at the Technion in Haifa is papered with citations: the Israel Prize; the Wolf Prize; and, in October 2011, the Nobel Prize in Chemistry.

Your Majesties, Your Royal Highnesses, Ladies and Gentlemen, the award ceremony speech in Stockholm began. "Having the courage to believe in his observations and in himself, Dan Shechtman has changed our view of what order is and has reminded us of the importance of balance between preservation and renewal, even for the most well-established paradigms…The disbelief that met Dan Shechtman was appropriate and healthy. Questioning should be mutual to promote the growth of knowledge. The ridicule he suffered was, however, deeply unfair. It is far too easy for all of us to remain in our lofty positions, and with lofty disdain regard the fool who claims that we are all wrong. To be that fool on the ground takes great courage, and both he and those that spoke out on his behalf deserve great respect."[4]

Marjorie Seneschal [*sic*] wants to make this book a kaleidoscope, Dorothy wrote in a note she never showed me. Not that book, Dorothy, this one. Kaleidoscopic, may David Brewster rest in peace. Six turns of the wheel, similar pieces in ever-new configurations.

In their 70s, Dorothy and Dora grew close again. Life had pulled them in different directions, ground them down in different ways, but they kept their fighting spirits. Dora had returned to Carn Voel, the seaside home in Porthcurno the Russells had bought when their son John was small. John, a chronic schizophrenic, lived with her now. "Nevertheless she battled on," says her daughter Harriet Ward, "bombarding newspapers, attending committees and conferences to promote this or that cause, from abortion law reform to environmental conservation to rationalist education—and above all the campaign for nuclear disarmament."[5]

Did Dorothy ("No," she wrote, "can't be Dot at this time of life!!") confide her own trials to her oldest friend? There is no hint of them in her letters to Dora. Yet she wrote, "There has been nothing more beautiful in my life than those days with you and Bertie at Porthcurno."[6]

CAST OF CHARACTERS

Dorothy Wrinch—herein DW—interacted with so many characters that a reader can easily lose track of them. To minimize that chance, I list here those who played a significant role in her life or appear in these pages more than once, or both. These brief descriptions convey only the subject's relation to DW; for more about their lives and accomplishments, see the *Dictionary of Scientific Biography*, the *Dictionary of National Biography*, the Nobel Foundation website http://www.Nobelprize.org, and (in many cases) their own full-length biographies and autobiographies.

Gladys Anslow (1892–1969) joined the Smith College physics department in 1914 and received her Ph.D.from Yale University 10 years later. She was DW's closest friend on the Smith College faculty and her scientific collaborator for 20 years.

Elizabeth von Arnim (1866–1941), a novelist, married Francis, the second Earl Russell, in 1916. DW met her when she studied with his brother Bertrand in the Earl's home in London. She was Elizabeth's frequent guest at Telegraph House, her country home.

William T. Astbury (1898–1961) pioneered the study of fibrous proteins (such as hair, silk, wool, keratin) by x-ray diffraction. An early supporter of DW's cyclol theory, he became a vociferous opponent.

Hertha Ayrton (1854–1923) was one of Girton College's first students, a founder of its Mathematical Club, and the first woman elected to the Institution of Electrical Engineers. DW held a Hertha Ayrton Fellowship in the early 1930s.

John Desmond Bernal (1901–1971) was a polymathic x-ray crystallographer known for his wide-ranging, brilliant intellect and fertile ideas. Bernal and DW were friends in their Theoretical Biology Club years, but enemies later.

Dora Black (see Dora Russell).

Carla Blake (1927–), second daughter of DW's sister Muriel.

Katharine Blodgett (1898–1979) was the first woman to earn a physics doctorate from Cambridge University and the first to join the General Electric research lab, where she worked with Irving Langmuir. She invented "invisible glass" for lenses, screens, and windshields.

Niels Bohr (1885–1962), a physicist and founder of quantum mechanics, had a deep interest in biology too. He saw DW's protein model as a possible link between biology and physics. Bohr won the Nobel Prize in Physics in 1922.

Andrew Donald Booth (1918–2009) was a pioneer of modern computing. Booth wrote several papers with DW in the 1940s on Fourier transforms and x-ray diffraction.

William Henry Bragg (1862–1942) received the Nobel Prize in Physics in 1915, jointly with his son **William Lawrence Bragg (1890–1971),** for using x-ray diffraction to determine the atomic structure of salt, diamond, and other simple mineral crystals.

Martin Buerger (1903–1986), a mineralogist and x-ray crystallographer at MIT, developed DW's interpretation and solution of Patterson maps into a widely used method for determining crystal structures.

Francis Crew (1886–1973) was a Scottish geneticist, friend of DW in the 1930s.

James G. Crowther (1899–1983), a friend of DW, was the Manchester *Guardian*'s first science writer.

Raphael Demos (1891–1968) met and impressed Bertrand Russell while a philosophy student at Harvard. Demos and DW were close during Russell's prison period and for a short while afterward.

Arthur Stanley Eddington (1882–1944) was a leading astrophysicist and friend of DW's first husband, John Nicholson.

Isidor Fankuchen (1905–1964), an American crystallographer, known as "Fan," was an early critic of DW's cyclol theory but became a friend and supporter.

Sara Margery Fry (1874–1958) attended Somerville College and served as its principal from 1926 to 1930. Fry was Pamela Wrinch's "godless godmother" and DW's lifelong confidante.

Otto Charles Glaser (1880–1951) was an Amherst college biologist and DW's second husband.

James Gray (1891–1975), a Cambridge biologist and author of *Experimental Cytology*, encouraged DW to study cell division in the early 1930s.

Lynda Grier (1880–1967) was headmistress of Lady Margaret Hall, Oxford, when DW taught there.

John Scott Haldane (1860–1936) was an eminent physiologist and a sophisticated supporter of "vitalism." His debate with D'Arcy Thompson in 1918 echoes throughout this book.

Godfrey Harold Hardy (1877–1947) was a leading Cambridge (later Oxford) mathematician and one of DW's key mentors.

David Harker (1906–1991), American x-ray crystallographer and student of Linus Pauling, was the first American to solve a protein structure. Harker became interested in proteins through conversations with DW in the 1930s.

Dorothy Crowfoot Hodgkin (1910–1994), a protein x-ray crystallographer, won the Nobel Prize in Chemistry in 1964 for solving the structure of vitamin B_{12}. She and DW clashed over the structure of insulin in the late 1930s; Hodgkin solved it in 1969.

Harold Jeffreys (1891–1989), a geophysicist, helped revive Bayesian probability theory in the twentieth century. The papers on scientific method he wrote with DW between 1919 and 1923 were the foundation of this and much of his later work.

Katharine Jex-Blake (1860–1951) was the headmistress of Girton College from 1916 to 1922, the period of DW's graduate study and Yarrow Research Fellowship.

Rosalie Johnson (1930–) is the youngest daughter of DW's sister Muriel.

W. E. Johnson (1858–1931) was a logician at King's College, Cambridge, influential in the development of probability theory. DW studied with him from 1916 to 17.

E. E. Constance Jones (1848–1922) was Girton's headmistress from 1903–1916.

Owen Jones (1809–1874) designed the exhibition cases for the Crystal Palace and later wrote *The Grammar of Ornament*, a compendium of symmetrical patterns and a touchstone for this book.

Stanley King (1883–1951) was president of Amherst College from 1932 to 1946.

Heinrich Klüver (1897–1979) was a neurophysiologist and expert on the physiological effects of the drug mescaline.

Irving Langmuir (1881–1957), DW's foremost supporter, won the Nobel Prize in Chemistry in 1934. In 1939, he nominated her for same.

A. E. H. Love (1863–1940), an Oxford mathematician, predicted a new type of the seismic waves. DW and Harold Jeffreys showed, in 1923, that his prediction was correct, and named them Love waves. Love served as a judge for DW's Oxford D.Sc.

Alan Mackay (1926–), British crystallographer and student of Bernal, was the first to surmise that a nonperiodic pattern, such as a Penrose tiling, might diffract waves as a crystal does.

Constance Malleson (1895–1975), an actress better known (and always called) by her stage name, Colette O'Neil, befriended DW in 1918 while her lover and DW's mentor, Bertrand Russell, was in jail.

Karl Menger (1902–1985) was a Viennese mathematician. DW met Menger in Vienna in 1932.

Ottoline Morrell (1873–1938) was Bertrand Russell's lover from 1911 to 1917 and a close friend thereafter. Their voluminous correspondence is a major source of information on DW's life between 1916 and 1918.

Hermann Muller (1890–1967), an American geneticist, was a member of the biology department at Amherst College from 1940 to 1945. He won the Nobel Prize in Physiology or Medicine the next year.

Joseph Needham (1900–1995), a biochemist and embryologist at Cambridge University, was a founding member of the Theoretical Biology Club. After World War II, he devoted his life to the history of Chinese science.

Dorothy Moyle Needham (1896–1987) was a biochemist like her husband and a lifelong friend of DW. She was one of the first women elected to the Royal Society of London.

John von Neumann (1903–1957), a Hungarian mathematician and architect of the modern computer, invited DW to work with him on deciphering protein structure from x-ray data, but such massive computations were not feasible at that time.

Eric Neville (1889–1961), a British mathematician at the University of Reading, was DW's lover and closest confidant for thirty years.

John Nicholson (1881–1955), a British mathematical physicist, supervised DW's M.Sc. and D.Sc. theses at the University of London. They were married in 1922 and divorced in 1938.

Carl Niemann (1908–1964) was an outspoken opponent of DW's protein model. With Linus Pauling he wrote the paper known as "The Debunking of Wrinch."

Charles Kay Ogden (1889–1957), publisher, inventor of Basic English, and admirer of Jeremy Bentham, was a friend of DW in her Cambridge years. Ogden—DW called him Og—published the first English-language edition of Wittgenstein's *Tractatus*.

Colette O'Neil (see Constance Malleson).

Arthur Lindo Patterson (1902–1966) developed "Patterson maps" for interpreting x-ray diffraction data. DW made important contributions to this method.

Linus Pauling (1901–1994) won the Nobel Prize in Chemistry in 1954 and for Peace in 1962; he was DW's foremost and most influential critic.

Michael Polanyi (1891–1976) was a Hungarian chemist and later a philosopher.

Alice Maud Procter (1862–1946) was the headmistress of Surbiton High School from 1899 to 1934. Devout, strict, and a "determined feminist," she directed DW's education from kindergarten through secondary school.

Robert Robinson (1886–1975) won the Nobel Prize in Chemistry in 1947. His experiments testing DW's protein model did not support it.

Bertrand Russell (1872–1970) was one of the century's foremost philosophers and public intellectuals. DW was his student and occasional assistant. Russell won the Nobel Prize for Literature in 1950.

Dora Russell (1894–1986), nee Black, was DW's best friend at Girton College. DW introduced her to Bertrand Russell, whom she married in 1921.

Francis (Frank) Russell (1865–1933) was Bertrand Russell's older brother and the second Earl Russell.

Leopold Růžička (1887–1976) won the Nobel Prize in Chemistry in 1939. DW consulted him on details of her cyclol model.

Dan Shechtman (1941–) received the Nobel Prize in Chemistry in 2011 for the discovery of quasicrystals.

Charles Percy Snow (1905–1980), a novelist remembered for his 1959 lecture on "The Two Cultures," began his career as a physicist. His first novel, *The Search*, is loosely based on the efforts of the Theoretical Biology Club to start a research institute.

Harry Sobotka (1899–1965) was a chemist at Mount Sinai Hospital in New York and a good friend of DW.

Milton Soffer (1914–1985) taught chemistry at Smith College for 43 years. It was he who told DW that her cyclol rings had been found.

Arthur Stoll (1887–1971) directed the Sandoz Laboratories in Basel. He found DW's cyclol rings in ergot alkaloids about 1950.

Bertha Swirles Jeffreys (1903–1999) studied mathematics at Girton and later taught there. She married DW's former collaborator Harold Jeffreys in 1945.

D'Arcy Thompson (1860–1948), naturalist and classicist, and author of *On Growth and Form*, urged mathematicians and physicists to study biology problems. After Russell, he was DW's most influential mentor.

Olga Taussky Todd (1906–1995), a mathematician, knew DW in Vienna (1931) and afterward.

Harold Urey (1893–1981) won the Nobel Prize in Chemistry in 1934 for the discovery of heavy hydrogen. Urey supported DW in the cyclol controversy.

Conrad Waddington (1905–1975), a geneticist, was a founding member of the Theoretical Biology Club.

George Neville Watson (1886–1965), a mathematician, directed DW's mathematical studies at Girton and helped her land her first job.

Warren Weaver (1894–1978), director of the Natural Sciences Division of the Rockefeller Foundation from 1932 to 1955, urged mathematicians and physicists toward biology. His interactions with DW are a major strand in this book.

Norbert Wiener (1894–1964) was a leading mathematician of the twentieth century. Wiener provided Patterson with his key insight.

Talitha Williams (1926–) is the oldest daughter of DW's sister Muriel.

Ludwig Wittgenstein (1889–1951) was a philosopher known best for his first book, *Tractatus Logico-Philosophocus*; DW secured its first (German) publisher.

Joseph Woodger (1894–1981), a biologist-turned-logician, was a founder of the Theoretical Biology Club.

Ada Minnie Souter Wrinch (1867–1934) was DW's mother.

Hugh Edward Hartr Wrinch (1866–1934) was DW's father.

Muriel Wrinch (1900–1965), DW's younger sister, married Helmut Schulz, with whom she had five children. She spent most of her adult life in South Africa.

Pamela Wrinch (1927–1975) was the daughter of DW and John Nicholson. A political scientist, Pamela taught in Boston colleges. She married Alfred Schenkman in 1967.

JACQUES HADAMARD'S ASSESSMENT OF DOROTHY WRINCH'S MATHEMATICAL WORK

December 20, 1930
Ecole Polytechnique
21 Rue Descartes, Paris V^e

It is with the greatest pleasure that I give my assessment of the work of Mrs. Wrinch.

Mrs. Wrinch has, from the beginning of her career, studied some of the most delicate problems in set theory and, consequently, in the philosophy of mathematics. Thus in 1923 she began to study the possibility of a cardinality intermediate between denumerable [the integers] and the continuum [the irrational numbers].[1] She was not able to decide whether this intermediate exists—this major problem is still unsolved despite a half century of effort: but her work will have been a useful contribution to this problem. Many other works show it to be related to the most delicate questions raised by contemporary mathematical logic.

At the same time, Mrs. Wrinch has dedicated herself to more concrete problems and, which is praiseworthy, is drawn to questions in mathematical physics and, in first place, by boundary value problems. These questions, which will without doubt be studied as long as there are mathematicians, are very difficult. In particular, the role played by the shape of the boundary which carries the values is mysterious. Except for the magnificent but unfortunately difficult to handle discoveries of Fredholm, it is only for very exceptional forms of this boundary that the geometers of the XIXth century supply the solution.

Mrs. Wrinch has succeeded in handling the most important of boundary problems for remarkably general domains by a skillful use of trigonometric series. She has studied domains in two dimensions, but the method is sufficiently general to extend to the case of surfaces and solids of revolution, passing from a line to the surface that it generates by turning about a fixed axis.

One particularly interesting feature of her method is that the case of a boundary presenting a cusp is not excluded. This case, so important to aviation,

---

Copies of the original handwritten letter, and a typescript, both in French, are in the Wrinch papers. This is my translation.—MS

preoccupied such researchers as Hopf and Villat, who considered it many times. Her method gives Mrs. Wrinch simple and malleable forms of the solution, applied to figures in the categories we are speaking of. Mrs. Wrinch has continued to study this subject and at the Congress of Mathematicians in Bologna (1928) presented views destined, it would seem, to be fruitful. There is no doubt that there will be, in the future, results worthy of interest.

One should not, however, omit another aspect of her activity, which is philosophy. In addition to problems in the theory of sets, as we have seen, she has been concerned from the first, and is still concerned, with conditions of scientific research and the scientific method. We have had the occasion, at the Congress of Philosophy a few years ago in Paris, to appreciate the ingenuity and range of her views. The reflections that the new and audacious conceptions of modern physics have suggested to her do not merit less attention.

Such a set of publications shows great penetration and a rare openness of spirit, a condition important for fruitful labor in the future.

If I cannot speak of the pedagogical and professional role of Mrs. Wrinch, at least I have been able, at Paris and Bologna, to appreciate the clarity of her expositions on profoundly different subjects, philosophy and the most concrete mathematics. But in any case, as concerns her work and scientific creations, I can associate myself without reservation with the opinion of my colleague M. Love.

J. Hadamard,
Member of the Académie des Sciences

ACKNOWLEDGMENTS

Archivists are the unsung heroes of scholarship. Let me sing the praises of the many who helped me over the years I worked on this book. At Smith College, Nanci Young and Karen Kukil answered questions and rekindled my enthusiasm again and again. Tom Rosenbaum and Bethany Antos at the Rockefeller Archive Center provided a wealth of materials over several visits, and solved strange mysteries. Ken Blackwell and Sheila Turcon at the Bertrand Russell Archive guided me through a vast trove of letters and helped me understand Wrinch's relationship with Russell and those close to him. Kate Perry at Girton College, Cambridge, and Pauline Adams at Somerville College, Oxford, tracked down records, letters, council minutes, and more. John Jones at Balliol College, Oxford, annotated the College correspondence concerning John Nicholson's illness and made it available to me. Verity Andrews at the University of Reading guided me through Eric Neville's papers, and Moira Mackenzie at St. Andrews University sent me D'Arcy Thompson's extensive correspondence with Dorothy Wrinch. I would also like to thank Jonathan Harrison and librarian Kathryn McKee at St. John's College, Cambridge; Richard Temple at the University of London; Tiny de Boer at the Internationaal Instituut voor Sociale Geschiedneis, Amsterdam; Janet Foster of the London Mathematical Society; Nichola Court at the Royal Society; Rose Lock at Sussex University; Mariah Sakrejda-Leavitt at Amherst College, and archivists of St. Hilda's and Lady Margaret Hall, Oxford; the Royal Institution, London; the Bodleian Library, Oxford; Cambridge University Library; the Library of Congress, the American Institute of Physics, and Yale University.

Anne and David Mininberg in Ossining, New York, and Paul Levy and Penny Marcus in Long Hanborough, Oxfordshire, gracious hosts, great cooks, archive mavens, and stimulating companions, made the Rockefeller and Oxford archives vacation destinations.

I am grateful to members of the widely scattered Wrinch family—Carla Blake, Claire Bohm Blake and Karl-Heinz Bohm, Rosalie Byrne, Rosalie Johnson, Mark

R. Schulz, Christopher Wrinch-Schulz, David Wrinch-Schulz, Julian and Mami Williams, and Talitha Williams for their memories, hospitality, photographs, and their interest in their remarkable relative and in bringing her story to light. I acknowledge with gratitude their permission to quote from her letters and papers in the many archives where they lie scattered.

I am grateful to Facebook, without which I might never have found the Wrinches. (But no, I will not create a Dorothy Wrinch Facebook page. For more about Dorothy, including the full list of her publications, see my website, http://www.marjoriesenechal.com.)

Others who knew Dorothy Wrinch or her close associates and shared their memories include Franz Alt, Maria Banerjee, Andrew Donald Booth, Kathleen Britten Booth, John Burk, Lale Burk, Carolyn Cohen, George Fleck, Helen Haddad, Luke Hodgkin, Elizabeth Horner, William Lipscomb, Alan Mackay, Lady Muir Wood, John Todd, and Olga Taussky-Todd. I would also like to thank Luke Hodgkin for permission to quote the first, handwritten versions of his mother's comments on Dorothy Wrinch.

Over years of lunches and long walks, Joan Afferica, Phyllis Cassidy, Marian Macdonald, and Marilyn Schwinn Smith listened patiently to my struggles to make sense of this convoluted story. Marian gamely slogged with me from Guildford to Shere in fiercely driving rain, retracing the walking tour I describe in the prologue. It is a pleasure to thank them and also Kristina Closser, Carolyn Cohen, Natalie Davis, Ivor Grattan-Guiness, June Barrow-Green, Istvan Hargittai, John Jones, Steven Lipson, Liz Louden, Alan Mackay, Revan Schendler, Berthold Schweizer, David Smith, Gary Werskey, Lady Muir Wood, and Greg Young for materials, photographs, and putting me in touch with helpful people and books. My special thanks to Sibilla Kennedy for making me aware of the vast resources of the Rockefeller Foundation Archives, to Alan Mackay for pointing out the importance of Wrinch's work in diffraction theory, to John Jones for sending me the correspondence in Balliol's archives concerning John Nicholson's breakdown and for organizing it so helpfully, and to Helen Weaver for the long-hidden key to the mystery. "What really happened to Dorothy Wrinch?"

The Fly Swatter: How My Grandfather Made His Way in the World, by Nicholas Dawidoff, and *The Lost: A Search for Six of Six Million*, by Daniel Mendelsohn, showed me how to walk a participant-investigator tightrope through complex stories of complex people in turbulent times. Creative writing workshops at the Banff International Research Station and multidisciplinary research projects at the Kahn Institute at Smith helped me hone that balancing act.

Andrea Barrett, Ken Blackwell, John Burk, Lale Burk, Nancy Campbell, Patrick Coffey, George Fleck, Evangeline Garreau, Barry Goldstein, Ivor Grattan Guinness, June Barrow-Green, Istvan Hargittai, Rene Heavlow, Kathleen Imholz, John Jones, Evelyn Fox Keller, Marian MacDonald, Alan Mackay, Edwin Mares, Judith Mindlin, Bill Oram, Stan Sherer, Sheila Turcon, Daniel Wikler and Jeanne Wikler read or listened to versions of the manuscript; I am deeply grateful for their criticisms and suggestions. Errors undoubtedly remain but, dear reader, without their expert intervention there would be a great many more.

Evangeline Garreau helped compile the cast of characters and ascertained the exact boxes and folders that now house Dorothy's letters at Smith.

And finally, my very deep thanks to my friend and writer-model Andrea Barrett, my friend and agent Regula Noetzli, and my friend and husband Stan Sherer for believing in the importance of Dorothy Wrinch's story and in my ability to tell it; and to my editor, Jeremy Lewis, for concurring.

This book is for Dorothy's friend and mine, Carolyn Cohen, who urged me to write it for 35 years. Scientist, humanist, and human being extraordinaire, Carolyn was inspired by Dorothy to make proteins her life work. She shares Dorothy's passion for the Beauty of science, but has never confused it with Truth.

REFERENCES AND NOTES

REFERENCES

I. Abbreviations for Special Collections

BAL Balliol College Archives, Oxford.

BRA Bertrand Russell Archives, McMaster University, Hamilton.

CKO C. K. Ogden Papers, William Ready Division of Archives and Research Collections, McMaster University Library.

DW Dorothy Wrinch Papers, Sophia Smith Collection, Smith College, Northampton.

DT D'Arcy Thompson Papers, University of St. Andrews Library, St. Andrews.

GCAR Girton College Archives, Cambridge.

IF Isidor Fankuchen Papers, 1933–1964, American Institute of Physics, Niels Bohr Library & Archives, College Park.

JDB J. D. Bernal Papers, Cambridge University Library.

JN Joseph Needham Papers, Cambridge University Library.

LMH Archives of Lady Margaret Hall, Oxford.

LP *Linus Pauling Day-by-Day*, Special Collections, Oregon State University, http://osulibrary.oregonstate.edu/specialcollections/coll/pauling/calendar.

RAC Rockefeller Archive Center, Sleepy Hollow, New York.

SC Somerville College Archives, Oxford.

WLB William Lawrence Bragg Papers, Royal Institution, London.

I am grateful to the Syndics of Cambridge University for permission to quote from the Bernal and Needham papers; to the Mistress and Fellows of Girton College for permission to quote from their archives; and to the Principal and Fellows of Lady Margaret Hall, Oxford, for permission to quote from the Grier papers.

II. Touchstone Books (cited extensively)

Ayling, Jean. [Dorothy Wrinch]. *The Retreat from Parenthood*. Kegan Paul, Trench, Trubner, 1930.
Cold Spring Harbor Symposia on Quantitative Biology. Vol. 6. The Biological Laboratory, 1938.
Jones, Owen. *The Grammar of Ornament*. Bernard Quartich, 1968.
Thompson, D'Arcy. *On Growth and Form*, Cambridge University Press, 1917.

NOTES

CHAPTER 1

1. Dora Russell, *The Tamarisk Tree: My Quest for Liberty and Love* (G. P. Putnam's Sons, 1975).

CHAPTER 2

1. "The molecular vision of life" is the title of a book by Lily Kay, subtitled: "Caltech, The Rockefeller Foundation, and the Rise of the New Biology" (Oxford University Press, 1993). "Molecular biology" was coined by Warren Weaver in 1938; see Warren Weaver, "Molecular Biology: the Origin of the Term," *Science* 170, 3958 (November 6, 1970), 581–582.
2. Arthur Lesk, "The Unreasonable Effectiveness of Mathematics in Molecular Biology," *Mathematical Intelligencer* 22, no. 2 (2000), 28–37.
3. Warren Weaver diary excerpt, Monday, June 15, 1936, Folder 498, Box 38, series 401D, Record Group (RG) 1.1, RAC.
4. David Ruelle, *The Mathematician's Brain* (Princeton University Press, 2007).
5. For the context of the discovery and the award, see Lily E. Kay, "W. M. Stanley's Crystallization of the Tobacco Mosaic Virus, 1930–1940," *Isis* 77, no. 3 (1986), 450–472.

CHAPTER 3

1. In 2011, Smith College replaced these windows with ones that can, with some effort, be opened.
2. William Waterhouse, "The Discovery of the Regular Solids," *Archive for History of Exact Sciences* 9, no. 3, 212–221.
3. E. H. Gombrich, *The Sense of Order: A Study in the Psychology of Decorative Art* (Cornell University Press, 1984).
4. M. Senechal and G. Fleck, eds., *Patterns of Symmetry* (University of Massachusetts Press, 1977).

CHAPTER 4

1. In September 1893, Ada Minnie Souter boarded the Royal Mail Ship *Trent* in Southampton, bound for South America.
2. Woodbridge, Suffolk, England website, http://woodbridgesuffolk.info/Woodbridge/history.htm.
3. From a pamphlet outlining the career of Dorothy's father, H. E. H. Wrinch, DW Box 1, Folder 8.
4. "From a Railway Carriage," Robert Louis Stevenson, *A Child's Garden of Verses* (1885).
5. David Rock, *Argentina: 1516–1987: From Spanish Colonization to Alfonsín* (University of California Press, 1987).
6. http://freepages.genealogy.rootsweb.ancestry.com/~tiber/Extracts.html. Hosted by rootsweb, an Ancestry.com community.
7. http://www.argbrit.org/SanBart/marrs1890–94.htm/. See the website British Settlers in Argentina—studies in 19th- and 20th-century emigration.
8. W. H. Shrubsole, *Where to Live Round London (Southern Side)* (Homeland Assoc., Ltd., 1905). Available as a Google e-book.
9. Virginia Woolf, *The London Scene: Six Essays on London Life* (Ecco Anniversary 25, 1975). These essays were originally published in the British edition of *Good Housekeeping* in 1931 and 1932.
10. W. L. Read and K. A. Knell, "Working with Steam Duties Performed by Men Working at the Chaddar's Lane Pumping Station 1894–1968," Cambridge Museum of Technology, 1996.
11. Charles Kingsley, "A Farewell," 1856.
12. *The Jubilee Book of Surbiton High School* (Favil Press, 1934).
13. I have adapted much of my account of the Procter family from Alice Procter's sister Zoë's autobiography, *Life and Yesterday* (Favil Press, 1960).
14. A writer for *The New Yorker* spun an article around her fruitless search for the author of that line. Rebecca Mead, "Middlemarch and Me," *The New Yorker*, February 14 and 21, 2011.
15. Alfred Lord Tennyson, "The Princess," in *The Princess: A Medley*, 1847.
16. Barbara Leigh Smith Bodichon, a cousin of Florence Nightingale. Barbara Bodichon is said to have been her friend George Eliot's model for *Romola*.
17. http://en.wikipedia.org/wiki/Girls'_Day_School_Trust.
18. Lytton Strachey, *Eminent Victorians* (Chatto and Windus, 1918), 168.
19. Douglas, *Jubilee Book*, 20.
20. Dorothy Wrinch, personal notes, n.d., DW Box 21, Notebook 21a.
21. Eliza Burney to Girton College, April 17, 1912, GCAR 2/4/1/16.

CHAPTER 5

1 E. E. Constance Jones, *Girton College*, Beautiful Britain Series (London, 1913).

2. M. C. Bradbrook, *"That Infidel Place": A Short History of Girton College 1869–1969* (Chatto and Windus, 1969).

3. Ibid.

4. Apostolos Doxiadis and Christos Papadopulous, *Logicomix* (Bloomsbury, 2009).

5. P. Sargant Florence, "Cambridge 1909–1919 and Its Aftermath," in *C. K. Ogden, A Collective Memoir*, ed. P. Sargant Florence and J. R. L. Anderson, 13–55 (Pemberton, 1977).

6. Evelyn Sharpe, *Hertha Ayrton: A Memoir* (Edward Arnold, 1926).

7. Nicholas Griffin, ed., *The Selected Letters of Bertrand Russell: The Public Years, 1914–1970* (Routledge, 2001), 243.

8. Ibid., 279.

9. C. P. Snow, *Variety of Men* (Charles Scribner's Sons, 1966).

10. William Whewell, *Of a Liberal Education in General; and with Particular Reference to the Leading Studies of the University of Cambridge* (J. W. Parker, 1845), 30.

11. Quoted by Andrew Brown in *J. D. Bernal, the Sage of Science* (Oxford University Press, 2005), 33.

12. For more on this journal, see I. Grattan-Guinness, "The Hon. Bertrand Russell and *The Educational Times*," in *Russell, Journal of Bertrand Russell Studies* 11, 1991, 86–91.

13. Pearson published a few papers with Cave's name on them too. But, she complained, she did calculations only and never saw the big picture.

14. Dot grasped this difficult material so well that Watson asked her to proofread the second, 1920, edition.

15. The field Cambridge then called the Moral Sciences has since splintered into logic, philosophy, psychology, and economics.

16. John G. Slater, ed., *Bertrand Russell: The Philosophy of Logical Atomism and Other Essays, 1914–1919* (George Allen and Unwin, 1986), 28.

17. Norbert Wiener, *Ex-Prodigy* (MIT Press, 1966).

18. Dorothy Wrinch to Bertrand Russell, September 9, 1914, BRA Collection RA1 710.057973. Russell's reply is lost.

19. *Cambridge Magazine*, October 16, 1915.

20. Eileen Rubery and Deryn Watson, "Girtonians and the World Wars," *Girton Project Journal*, 1, Girton College (April 2009), 12.

21. Dorothy Wrinch to Karl Pearson, May 12, 1916, Pearson Papers, University College, London Library.

22. Dorothy Wrinch to Bertrand Russell, May 12, 1916, BRA Collection RA1 710.057975.

CHAPTER 6

1. Bertrand Russell to Ottoline Morrell, June 8, 1916, *Selected Letters*, 68.
2. Dorothy Wrinch to C. K. Ogden, July 8, 1916, CKO.
3. G. H. Hardy, *Bertrand Russell & Trinity* (Cambridge University Press, 1942; reprinted Arno Press, 1977).
4. http://en.wikipedia.org/wiki/Frank_Russell,_2nd_Earl_Russell.
5. Bertrand Russell to Ottoline Morrell, quoted by Ray Monk, *Bertrand Russell: The Spirit of Solitude, 1872–1921* (Free Press, 1996), 351.
6. T. S. Eliot, "Mr. Apollinax." See also Donald J. Childs, "Mr. Apollinax, Professor Channing-Cheetah, and T. S. Eliot," *Journal of Modern Literature* 13, no. 1 (May 1986), 172–177.
7. Victor F. Lenzen, "Bertrand Russell at Harvard, 1914," *Russell: The Journal of Bertrand Russell Studies* 91, 1971. Available at http://digitalcommons.mcmaster.ca/russelljournal/vol91/iss3/4.
8. G. H. Hardy, *A Mathematician's Apology* (Cambridge University Press, 1940).
9. Bertrand Russell and A. N. Whitehead, *Principia Mathematica*, vol. 1 (Cambridge University Press, 1910), 37.
10. Elizabeth Hardwick, "Bloomsbury and Virginia Woolf," in *Seduction and Betrayal: Women and Literature* (NYRB Classics, 2001), 124.
11. Marjorie Hope Nicholson, *Pepys' Diary and the New Science* (Charlottesville: University of Virginia Press, 1965).
12. Lenzen, "Bertrand Russell at Harvard, 1914."
13. Dorothy Wrinch to C. K. Ogden, n.d., December 1916 (?), CKO.
14. Snow, *Variety of Men.*
15. G. H. Hardy to Katharine Jex-Blake, July or August 1917, GCAR 2/4/1/16.
16. Bertrand Russell to Girton College, March 17, 1918, GCAR 2/4/1/16.
17. D. M. Wrinch, "Mr. Russell's *Lowell Lectures*," *Mind*, n.s., 26, no. 104 (1917).
18. Dorothy Wrinch, "On the Nature of Judgment," *Mind* 28, no. 3 (1919), 319–329; "On the Nature of Memory," *Mind* 29, no. 11 (1920), 46–61.
19. A. C. Grayling, *Russell: A Very Short Introduction* (Oxford University Press, 2002).
20. L. L. Whyte, *Essay on Atomism, from Democritus to 1960* (Wesleyan University Press, 1961; Harper-Row, 1963), 12.
21. *Stanford Encyclopedia of Philosophy*, http://plato.stadford.edu/entries/logical-atomism.
22. Edwin Mares, "Russell's Logical Forms," *Soochow Journal of Philosophical Studies*, no. 16 (2007), 215–256; Edwin Mares to MS, e-mail, October 19, 2011.
23. G. H. Hardy to Miss Clover, n.d., 1918, GCAR 2/5/5/28.

CHAPTER 7

1. Robert Gathorne-Hardy, ed., *Ottoline at Garsington: Memoirs of Lady Ottoline Morrel 1915–1918* (Alfred A. Knopf, 1975), 251.

2. Miles Malleson, an actor and screenwriter, appeared in over sixty British films, including *Kind Hearts and Coronets*, *The Captain's Paradise*, and *The Hound of the Baskervilles*.

3. Sheila Turcon, "Like a Shattered Vase: Russell's 1918 Prison Letters," *Russell: The Journal of Bertrand Russell Studies* 30 (Winter 2010–11), 101–125. Available at http://digitalcommons.mcmaster.ca/russelljournal/vol30/iss2/2.

4. Gladys Rinder to Bertrand Russell, June 21, 1918, BRA Collection RA1 710.054821.

5. Dorothy Wrinch to Bertrand Russell, n.d. [but soon after July 8, 1918], BRA Collection RA1 710.057983.

6. P. B. Medawar, "Postscript: D'Arcy Thompson and Growth and Form," in *D'Arcy Wentworth Thompson*, ed. Ruth Thompson (Oxford University Press, 1958).

7. *Oxford Dictionary of National Biography*, s.v. "Haldane, John Scott," by Steve Sturdy.

8. Ruth Thompson, quoted in *D'Arcy Wentworth Thompson*, 63–64.

9. Medawar, "Postscript," in *D'Arcy Wentworth Thompson*.

10. Joseph Needham, "Biological Aspects of Form and Growth," in *Aspects of Form*, ed. L. L. Whyte (Pellegrini and Cudahy, 1951).

11. Martin Goodman, *Suffer and Survive: The Extreme Life of John Scott Haldane* (Simon and Schuster, 2007).

12. See the *Proceedings of the Aristotelian Society*, n.s., 18 (1918), 419–461.

13. Poul Anderson, quoted in A. L. Mackay, *Scientific Quotations: Harvest of a Quiet Eye* (New York Institute of Physics, 1977).

14. Jeremy Bentham, *Introduction to the Principles of Morals and Legislation* (1789), chapter 4. "Felicific" is not a felicitous word; Bentham's calculus is sometimes called "hedonistic" or "emotional" instead.

15. Ibid.

16. Norbert Wiener, "Studies in Synthetic Logic," *Proceedings of the Cambridge Philosophical Society*, 18 (1915), 24–28.

17. G. E. Moore, *Principia Ethica*, 28. Quoted by Wrinch in "On the Summation of Pleasures."

18. Dana M. Small et al., "Changes in Brain Activity Related to Eating Chocolate, from Pleasure to Aversion," *Brain* 124 (2001), 1720–1733, available at http://brain.oxfordjournals.org/cgi/reprint/124/9/1720.

19. Daniel Wikler to MS, private communication, December 29, 2011.

20. Francis Jehl, *Menlo Park Reminiscences*, vol. 2 (Edison Institute, 1938), 482. Quoted in http://home.frognet.net/~ejcov/foxpitt.html.

21. Bertrand Russell to Frank Russell, July 15, 1918, BRA Collection RA1 079991; Frank Russell to Bertrand Russell, July 19, 1918, BRA Collection RA1 079993.

22. Bertrand Russel to Ottoline Morrell, August 14, 1918. Quoted in *Bertrand Russell: The Philosophy of Logical Atomism and Other Essays, 1914–19*, 249.

23. Lieut. Col. Newnham-Davis, "Dinners and Diners, Where and How to Dine in London" (Grant Richards, 1899), available at http://www.victorianlondon.org/publications2/dinners-30.htm.

24. Gladys Rinder to Bertrand Russell, June 21, 1918, BRA Collection RA1 710.054821.

25. Turcon, "Like a Shattered Vase."

26. Gladys Rinder to Bertrand Russell, shortly after July 8, 1918, BRA Collection RA1 710.054823.

27. Bertrand Russell to Constance Malleson, June 27, 1918, BRA Collection RA3 596.200312.

28. Constance Malleson to Bertrand Russell, July 10, 1918, BRA Collection RA3 596.104579GX.

29. Constance Malleson to Bertrand Russell, July 26, 1918, BRA Collection RA3 596.104579HC.

30. Constance Malleson to Bertrand Russell, July 28, 1918, BRA Collection RA3 596.104579HD.

31. Frank Russell to Bertrand Russell, July 19, 1918, BRA Collection RA1 730.079993.

32. Bertrand Russell to Frank Russell, July 22, 1918, BRA Collection RA1 730.079994.

33. Earl Russell, *My Life and Adventures* (Cassell, 1923), 324.

34. Constance Malleson to Bertrand Russell, August 19, 1918, BRA Collection RA3 596.104579HI.

35. Ottoline Morrell to Bertrand Russell, July 8, 1918, BRA Collection RA1 710.082673A; Bertrand Russell to Ottoline Morrel, July 14, 1918, BRA Collection RA1 710.053258.

36. Constance Malleson to Bertrand Russell, July 28, 1918, BRA Collection RA3 596.104579HD.

37. Bertrand Russell to Ottoline Morrell, August 1, 1918, BRA Collection RA3 385.001489F.

38. Griffin, ed., *Selected Letters*, 168.

39. See Katie Riophe, *Uncommon Arrangements: Seven Portraits of Married Life in Literary London Circles, 1910–1939* (Dial Press, 2007).

40. Bertrand Russell, *The Autobiography of Bertrand Russell, 1914–1944* (Allen and Unwin, 1968), 154.

41. Griffin, ed., *Selected Letters*, 169.

42. Constance Malleson to Bertrand Russell, August 25, 1918, BRA Collection RA3 596.079993A.
43. Dorothy Wrinch to Katharine Jex-Blake, August 18, 1918, GCAR 2/4/1/16.
44. Dorothy Wrinch to Bertrand Russell, July 18, 1919, BRA Collection RA1 710.057990.

CHAPTER 8

1. Russell, *The Tamarisk Tree*, 72.
2. Thomas Hardy, "At Lulworth Cove a Century Back," available on the Poetry Foundation website, http://www.poetryfoundation.org/poem/176677.
3. J. E. Littlewood, *A Mathematician's Miscellany* (Methuen, 1953); reprinted as *Littlewood's Miscellany*, B. Bollobás, ed. (Cambridge University Press, 1986), 128–129.
4. E. T. Bell, *The Handmaiden of the Sciences* (Williams and Wilkins, 1937).
5. Denis Overbye, "A Scientist Takes on Gravity," *New York Times*, July 12, 2010.
6. Bertrand Russell to Constance Malleson, Sept. 2, 1919, BRA Collection RA3 596.200534.
7. Bertrand Russell to Constance Malleson, Sept. 7, 1919, BRA Collection RA3 596.200359.
8. Russell, *The Tamarisk Tree*.
9. Katharine Tait, *My Father, Bertrand Russell* (1975; Thoemess Press, 1996).
10. Pnina Abir-Am, "Synergy or Clash: Disciplinary and Marital Strategies in the Career of Mathematical Biologist Dorothy Wrinch," in *Uneasy Careers and Intimate Lives, Women in Science 1789–1979*, ed. Pnina Abir-Am and Dorinda Outram (Rutgers University Press, 1987); hereafter "Synergy or Clash."
11. Dorothy Wrinch to Bertrand Russell, Sept. 25, 1919, BRA Collection RA1 710.057991.
12. "Meeting of the Royal Astronomical Society," *Observatory* 43, no. 548 (Jan. 1920), 33–45.
13. S. Chandrasekhar, *Truth and Beauty* (University of Chicago Press, 1987), 112–113.
14. *Oxford Dictionary of National Biography*, s.v. "Johnson, William Ernest," by R. B. Braithwaite.
15. Sir Alan Cook, *The Observational Foundations of Physics* (Cambridge University Press, 1994), 123.
16. David Howie, *Interpreting Probability: Controversies and Developments in the Early Twentieth Century* (Cambridge University Press, 2002), 31.
17. Dorothy Wrinch and Harold Jeffreys, "On Some Aspects of the Theory of Probability," *Philosophical Magazine*, sixth series, 38, no. 223 (July 1919), 715–731.

18. In the course of the twentieth century, inverse probability, better known as Bayesian, would rise, fall, and rise again, a saga recounted by Sharon Bertsch McGrayne in *The Theory That Would Not Die: How Bayes' Rule Cracked the Enigma Code, Hunted Down Russian Submarines, and Emerged Triumphant from Two Centuries of Controversy* (Yale University Press, 2011).

19. Minutes of the College Council, May 3, 1921, Somerville Archives, Somerville College, Oxford, UK.

20. http://en.wikipedia.org/wiki/Oppau_explosion.

21. A. E. H. Love, *Some Problems of Geodynamics: Being an Essay to Which the Adams Prize in the University of Cambridge Was Adjudged in 1911* (Cambridge University Press, 1911).

22. Dorothy Wrinch and Harold Jeffreys, "On the Seismic Waves from the Oppau Explosion of 1921 Sept. 21," *Geophysical Supplements to the Monthly Notices of the Royal Astronomical Society* 1, no. 2 (1923), 15–22.

23. F. M. Jaeger, *Lectures on the Principle of Symmetry* (Elsevier, 1920).

24. Dora Russell, "My Friend Ogden," in *C. K. Ogden: A Collective Memoir.*

25. Ivor Grattan-Guinness, *The Search for Mathematical Roots, 1870–1940* (Princeton University Press, 2000), 436.

26. The National Union of Scientific Workers was renamed the Association of Scientific Workers in 1927. By then, Harold Jeffreys had effectively dropped out; I do not know how long Dot remained active. For more about this important organization see Gary Werskey, *The Visible College* (Holt, Rinehart and Winston, 1978).

27. *The Times of London*, June 11, 1921.

28. Dorothy Wrinch and John Nicholson to Bertrand Russell, n.d., September 1920 (?), BRA Collection RA3 1027.250303.

29. Constance Malleson to Bertrand Russell, January 7, 1921, BRA Collection RA3 596.104579 LQ.

30. Frank Russell to Bertrand Russell, January 27, 1921. Quoted in *The Autobiography of Bertrand Russell, 1914–1944.*

31. P. Meisel and W. Kendrick, eds., *Bloomsbury/Freud: The Letters of James and Alix Strachey, 1924–1925* (Basic Books, New York, 1985), 223.

32. Harold Jeffreys, *Nature*, April 8, 1976, 564.

CHAPTER 9

1. P. Medawar in *D'Arcy Wentworth Thompson.*

2. See, for example, Susan Cain, "Shyness: Evolutionary Tactic?" *New York Times*, June 25, 2011.

3. "The Asymptotic Expansions of Solutions of Certain Differential Equations of the Third and Fourth Orders," for her master of science, M.Sc., and two papers, "An Asymptotic Formula for the Hypergeometric Function of $04+^{2-}$" and "A

Generalized Hypergeometric Function with n Parameters," for the doctor of science, or D.Sc.

4. A. Dendy, *Outlines of Evolutionary Biology*, 2nd ed. (D. Appleton, 1912), 420.

5. Thompson, *On Growth and Form*.

6. Aristotle, *History of Animals*.

7. J. W. Nicholson, "The Lateral Vibrations of Bars of Variable Section," *Proceedings of the Royal Society of London, Series A*, 93, no. 654 (1917), 506–519; A. Dendy and J. W. Nicholson, "On the Influence of Vibrations upon the Form of Certain Sponge-Spicules," *Proceedings of the Royal Society of London, Series B*, 89, no. 622 (Aug. 1, 1917), 573–587.

8. J. W. Nicholson, "Mathematics," in *Problems of Modern Science*, ed. A. Dendy (George G. Harrap), 1922.

9. Bradbrook, *That Infidel Place*, 65.

10. Dorothy Wrinch to Girton College, June 12, 1921, GCAR 2/4/1/16.

11. Mary Croaken, "Beautiful Numbers: The Rise and Decline of the British Association Mathematical Tables Committee, 1871–1965," *IEEE Annals of the History of Computing* 22, no. 4 (2000), 44–61. Croaken gives Dot's years of service as 1923–1929, but she joined in 1921.

12. H. E. H. Wrinch and Dorothy Wrinch, "Table of the Bessel Function $I_n(x)$," *Philosophical Magazine* 45 (1923), 846–849; "The Roots of the Hypergeometric Functions with a Numerator and Four Denominators," *Philosophical Magazine*, seventh series, 1 (1926), 273–276.

13. University of London Senate Minute 2695, July 1921.

14. *Dictionary of National Biography*, s.v. "Nicholson, John William," by William Wilson.

15. R. McCormmach, "The Atomic Theory of John William Nicholson," *Archive for History of Exact Sciences* 3, Springer-Verlag (1966/1967), 160–184.

16. John Nicholson to Bertrand Russell, n.d., 1922, BRA Collection RA3 250181.1027.28.

17. *The Manchester Guardian*, August 2, 1922, 14.

18. Bertha Swirles (Lady Jeffreys), Oral History, unpublished. The Mistress and Fellows of Girton College, Cambridge, GCAR GCPP Jeffreys 1/9/1.

19. Lynda Grier to Dorothy Wrinch, October 2, 1922, Grier papers, LMH.

20. John Nicholson to Lynda Grier, June 30, 1923, Grier papers, LMH.

21. For details of this arrangement, see Abir-Am, "Synergy or Clash."

22. Lynda Grier, November 17, 1930, DW Box 8, Folder 5.

23. D'Arcy Thompson to Dorothey Wrinch, September 15, 1924, DW, Box 5, Folder 23.

24. John's poignant tribute to the brilliant young mathematician was published in the *Proceedings of the London Mathematical Society*, series 2, 16 (1917). Maclaren had volunteered for military service. "We must admire such a

decision in one for whom the military life had no attraction, but at the same time regret a national state of mind which could not suggest, to a man of this type, that the best national work he could do was in a different category."

25. Dorothy Wrinch to D'Arcy Thompson, August 28, 1924, DT 24450.
26. D'Arcy Thompson to Dorothy Wrinch, September 15, 1924, DW, Box 5, Folder 23.

CHAPTER 10

1. Muriel Wrinch and H. H. Schulz, *Mothers and Babies: A Practical Book About the Everyday Life of the Baby from Birth to Four Years Old* (T. C. and E. C. Jack, 1924); Muriel Wrinch, *Mothers and Children* (Frederick A. Stokes, 1926).
2. Bertrand Russell to Dorothy Wrinch, January 29, 1920, BRA RA1 710.057997.
3. Abir-Am, in "Synergy or Clash," identifies S as Sybil Moholy-Nagy, perhaps because a letter in the Wrinch papers thanks Dot for the book. But that very formal letter is dated 1936. In 1930, the then Sybil Peech still worked in a film studio in her native Germany; the Moholy-Nagys moved to London in 1935.
4. Russell, *The Autobiography of Bertrand Russell, 1917–1944*, 153.
5. The university-appointed judges were Professor Love and G. B. Jeffery (a professor of mathematics at University College, London). Their handwritten report is in the records of the Bodleian Library.
6. D. M. Wrinch, "On the Asymptotic Evaluation of Functions Defined by Contour Integrals," *American Journal of Mathematics* 50 (1928), 269–302.
7. The letters quoted here can be found in DW Box 7, Folder 8.
8. Dorothy Wrinch, "On Harmonics Applicable to Surfaces of Revolution," *American Journal of Mathematics* 52 (1930), 305–318.
9. Bertha Swirles (Lady Jeffreys), GCAR GCPP Jeffreys 1/9/1.
10. Dot's letter of inquiry is lost. "Miss Clark," at Smith, forwarded it to Radcliffe College. Ada Comstock, then president of Radcliffe, wrote to Dot on April 2, 1929, explaining that there were no openings at Radcliffe nor was there likely to be any in the near future. DW Box 8, Folder 5.
11. Tim Weiner, "R. V. Jones, Science Trickster Who Foiled Nazis, Dies at 86," *New York Times*, December 19, 1997, http://www.nytimes.com/1997/12/19/us/r-v-jones-science-trickster-who-foiled-nazis-dies-at-86.html.
12. R. V. Jones to John Jones, February 11, 1993, BAL MBP 4 9*.
13. Collier to Duke, June 9, 1931, BAL MBP 49.ii.
14. A. Schwarz, "N.F.L. Acknowledges Long-Term Concussion Effects," *New York Times*, December 20, 2009.
15. A. D. Lindsay to The Visitor, November 8, 1930, BAL MBP 49.ii.a.

16. F. F. Urquhart, Vice Regent of Balliol, to The Official Solicitor, Royal Courts of Justice, February 17, 1931 BAL MBP 49.v.c.
17. Dora Russell to Bertrand Russell, July 17, 1930, BRA Collection RA2 710.104103.

CHAPTER 11

1. Dorothy Wrinch to Bertrand Russell, July 14, 1930, BRA Collection RA1 710. 057994.
2. Margery Fry, n.d., DW Box 4, Folder 31.
3. G. H. Hardy to Dorothy Wrinch, n.d., DW Box 8, Folder 5.
4. Reinhardt Siegmund-Schultze, *Rockefeller and the Internationalzation of Mathematics Between the Two World Wars* (Birkhäuser, 2001).
5. M. Senechal, "Hardy as Mentor," *The Mathematical Intelligencer* 29, no. 7 (2007), 16–23.
6. G. H. Hardy to Dorothy Wrinch, n.d., DW Box 8, Folder 5. After mathematics, Hardy's passion in life was cricket.
7. A. M. Carr-Saunders and P. A. Wilson, *The Professions* (Oxford at the Clarendon Press, 1933).
8. See correspondence between Bertrand Russell and W. W. Norton, BRA.
9. D'Arcy Thompson to the Academic Registrar, University of London, May 29, 1931, in support of Wrinch's application for a grant from the Dixon Fund. DT 24434.
10. Ibid.
11. Dorothy Wrinch to D'Arcy Thompson, June 2, 1931. DT 24435.
12. Quoted by Steindór Erlingsson in "The Rise of Experimental Zoology in Britain in the 1920s: Hogben, Huxley, Crew, and the Society for Experimental Biology" (Ph.D. thesis, University of Manchester, 2005), 82.
13. Quoted by Dorothy Wrinch in her undated, untitled draft of a proposal to the Leverhulme Foundation, typescript, JN.
14. Dorothy Wrinch to Lynda Grier, October 19, 1931, Grier papers, LMH.
15. The first explicit use of this term appears in Vladimir A. Kostitzin's *Biologie mathématique* (A. Colin, 1937). Vito Volterra used the term in his foreword to that book. (Kostitzin's book was published in English in 1939: *Mathematical Biology* [Harrap].) A. J. Lotka's *Elements of Physical Biology*, 1925, was renamed *Elements of Mathematical Biology* for the 1956 edition. (I am grateful to Professor Georgio Israel for this information.)
16. John Todd to Marjorie Senechal, private communication. October 20, 1997.
17. Karl Menger, *Reminiscences of the Vienna Circle and the Mathematical Colloquium*, vol. 20, ed. Louise Golland, Brian McGuinness, and Abe Sklar, Vienna Circle Collection (Kluwer Academic, 1994). Evidently, Professor Hahn did not cut his course short by announcing Menger's solution.

18. The University of California, Berkeley, hired Julia Robinson in 1976, after her election to the National Academy of Sciences. More honors followed: first woman president of the American Mathematical Society, a MacArthur "genius" award, and election to the American Academy of Arts and Sciences. Yet only Smith College gave her an honorary degree.

19. Sally Mitchell, *The New Girl: Girls; Culture in England, 1880–1915* (Columbia University Press, 1995).

20. *Jubilee Book*, 40.

21. Dorothy Wrinch to the Principal, Lady Margaret Hall, March 18, 1934, LMH.

22. Dorothy Wrinch, Leverhulme draft proposal, typescript, JN.

23. Dorothy Wrinch to J. G. Crowther, October 1934, JCG.

24. This shift in emphasis was, in part, a natural confluence of interests within the foundation, as its Medical Sciences division increasingly emphasized neurology, psychiatry, heredity and genetics, and virus research.

CHAPTER 12

1. Warren Weaver, *Scene of Change: A Lifetime in American Science* (Charles Scribner's Sons, 1970), 68.

2. Warren Weaver oral history #8, pp. 436–443, May 5, 1961, Vol. III, RG 13, RF Archives, RAC.

3. For more on British left-wing scientists in the 1930s, see *The Visible College*.

4. Joseph Woodger, "Some Problems of Biological Methodology," *Proceedings of the Aristotelian Society*, n.s., 29 (1928–29), 331–358.

5. Joseph Needham to Joseph Woodger, August 19, 1932, JN M98.

6. Dorothy Wrinch to Joseph Needham, September 10, 1932, JN J.243.

7. Dot described the paper (never published) in a letter to Bernal undated, summer 1932: "It's full of shells. Do you mind them having the foll. properties. 1. Considerable inertia. 2. The capacity to hold certain max. quantity of elect. after which any excess just goes away. 3. The surface is the set of surface energy and 4. Shells by their own metabolism can make electricity, so that when they expand (and so can hold more) they automatically charge up to the appropriate amount." JDB ADD 8287 J.258.

8. C. H. Waddington, *The Evolution of an Evolutionist* (Cornell University Press, 1975).

9. For a complete schedule of meetings and who attended them, see Pnina Abir-Am, "The Biotheoretical Gathering, Trans-disciplinary Authority and the Incipient Legitimation of Molecular Biology in the 1930s: New Perspective on the Historical Sociology of Science," *History of Science*, 25 (1987), 1–70.

10. Karl Popper, "Joseph Henry Woodger" *British Journal for the Philosophy of Science* 32, no. 3 (Sept. 1981), 328–330.

11. Karl Popper, *Conjectures and Refutation: The Growth of Scientific Knowledge* (Routledge and Kegan Paul, 1963).

12. Joseph Needham, *Order and Life* (Yale University Press, 1936).

13. Dorothy Wrinch to Joseph Needham, undated typescript, JN.

14. C. H. Waddington, *How Animals Develop* (W. W. Norton, 1936).

15. In the 1960s, the French topologist René Thom invented a mathematical theory for Waddington's ideas; he called it catastrophe theory.

16. J. M. W. Slack, "Conrad Hal Waddington: The Last Renaissance Biologist?" *Nature Reviews Genetics*, 3 (Nov. 2002), 889–895.

17. Francis Crick, *What Mad Pursuit* (Basic Books, 1988).

18. Dorothy Wrinch to Joseph Needham, February 23, 1935, JN B.22.

19. D. M. Wrinch, "On the Molecular Structure of Chromosomes," *Protoplasma* 25, no. 4 (1936), 550–569.

20. Delta thanked Dorothy Crowfoot (later Hodgkin) too, but she had never attended a meeting of the club. Later—in 1936 and 1937—Crowfoot attended two meetings at which Delta was not present. See Abir-Am, "The Biotheoretical Gathering."

21. Robert Olby to Dorothy Wrinch, n.d. (1970?), DW Box 5, Folder 7.

22. Robert Olby, *The Path to the Double Helix* (University of Washington Press, 1974; Dover, 1996).

23. *Nature*, Nov. 16, 1935, 786.

24. W. E. Tisdale, October 23, 1935, RG 12, RF Archives, RAC. A is W. T. Astbury; he is ruefully reflecting on his own reception.

CHAPTER 13

1. The remarks in this chapter, and many more, are quoted from W. E. Tisdale diary, RG 12 and Folders 498–500, Boxes 38–39, Series 401D, RG 1.1, RF Archives, RAC.

   B. is Henry Borsook, B.* is J. D. Bernal, Br. is William Lawrence Bragg, C. is Albert Charles Chibnall, C.* is E. J. Cohn, D. is Jack Cecil Drummond, E. is Jacques Errera, Et. is G. Ettisch, F. is Howard Florey, G. is E. Gorter, H. is A. V. Hill, H*v. is Emil Hatschek, H.** is Hans Clarke, Ha. is Charles Robert Harington, Ho. is T. R. Hogness, J. is Herbert Spencer Jennings, K. is Ralph Kekwick, L. is Charles Lovatt Evans, v. M. is Alexander Ludwig von Muralt, M.* is Herman Francis Mark, Me. is K. H. Meyer, P. is N. W. Pirie, P.* is [?] Peters, P.** is Linus Pauling, R. is Robert Robinson, S. is William Stanley, St. is Jacinto Steinhardt, U. is Harold Urey, Y. is J. Z. Young, Z. is Solomon Zuckerman.

CHAPTER 14

1. http://en.wikipedia.org/wiki/Orion_(constellation).
2. E. H. Gombrich, *Art and Illusion: A Study in the Psychology of Pictorial Representation* (Phaidon, 1960; Princeton University Press, 2000).
3. C. C. Furnas, *The Next Hundred Years: The Unfinished Business of Science* (Williams and Wilkins, 1936), 315.
4. Dorothy Wrinch to J. D. Bernal, May 7, 1934, JDB ADD 8287 J.258.
5. Warren Weaver diary, Monday, June 15, 1936, Folder 498, Box 38, Series 401D, RG 1.1, RF Archives, RAC.
6. Warren Weaver, diary, Sept. 11, 1936, Folder 498, Box 38, Series 401D, RG 1.1, RF Archives, RAC.
7. Donna Jeanne Haraway, *Crystals, Fabrics, and Fields: Metaphors That Shape Embryos* (Yale University Press, 1976; North Atlantic Books, Berkeley, 2004).
8. Albert Rosenfeld, "The Quintessence of Irving Langmuir," in *The Collected Works of Irving Langmuir*, vol. 12 (Pergamon Press, 1962).
9. I. Langmuir, "Fundamental Research and Its Human Value" (lecture to the Congress of Applied Chemistry, Paris, September 30, 1937). It appeared in the *General Electric Review* 40, no. 569 (1937) and in *The Collected Works of Irving Langmuir*, vol. 12.
10. Warren Weaver diary, December 22, 1936, Folder 498, Box 38, Series 401D, RG 1.1, RF Archives, RAC.
11. Finn Aaserud, *Redirecting Science: Niels Bohr, Philanthropy, and the Rise of Nuclear Physics* (Cambridge University Press, 1990).
12. Needham, *Order and Life*. 33.
13. Dorothy Wrinch to Niels Bohr, December 16, 1938, DW Box 4, Folder 29.
14. Waldemar Kaempffert, "Science in the News," *New York Times*, April 23, 1939.
15. Linus Pauling to Warren Weaver, March 6, 1937, Folder 499, Box 39, Series 401D, RG 1.1, RF Archives, RAC.
16. *New York Times*, September 11, 1937.
17. J. G. Crowther, *Fifty Years with Science* (Barrie and Jenkins, 1970).
18. Irving Langmuir, "The Structure of Proteins," *Proceedings of the Physical Society* 51, pt. 4, no. 286, 592 (July 1939); reprinted in *The Collected Works of Irving Langmuir*, vol. 7.
19. Elizabeth Crawford, *The Nobel Population, 1901–1950* (Universal Academic Press, 2002), 306.
20. Patrick Coffey, *Cathedrals of Science: The Personalities and Rivalries That Made Modern Chemistry* (Oxford University Press, 2008), 211.
21. Haraway, *Crystals, Fabrics, and Fields*.
22. Quoted by Abraham Pais, *Niels Bohr's Times, in Physics, Philosophy, and Polity* (Clarendon Press, 1991).
23. Ibid.

24. Thomas Hager, *Force of Nature: The Life of Linus Pauling* (Simon and Schuster, 1995), 99.

25. Georgina Ferry, *Dorothy Hodgkin: A Life* (Cold Spring Harbor Laboratory Press, 2000).

CHAPTER 15

1. Warren Weaver diary, Monday, June 15, 1936, Folder 498, Box 38, Series 401D, RG 1.1, RF Archives, RAC.

2. Dorothy Wrinch to W. E. Tisdale, January 10, 1936, DW Box 8, Folder 4.

3. Warren Weaver to W. E. Tisdale, January 24, 1936; reply same day, Folder 498, Box 38, Series 401D, RG 1.1, RF Archives, RAC.

4. Joseph Woodger to Joseph Needham, January 10, 1933, JN.

5. Walter Langford, typescript, E. H. Neville Papers, Reading University Library.

6. M. Senechal, "The Last Second Wrangler," in *The Shape of Content*, ed. C. Davis, M. Senechal, and J. Zwicky (A. K. Peters, 2007).

7. Reviewed by T. P. Nunn in the *Mathematical Gazette* 12, no. 168 (Jan. 1924), 27–30.

8. W. B. Yeats, "When You Are Old," *The Rose*, 1893.

9. Quoted in Susan Sherman, ed., *May Sarton: Collected Letters 1916–1954* (W. W. Norton, 1997).

10. George Sarton to May Sarton, June 29, 1939, Sarton Papers, Berg Collection, New York Public Library.

11. John Todd to MS, private communication, October 20, 1997.

12. W. E. Tisdale diary, March 14, 1936, RG 12, RF Archives, RAC.

13. Warren Weaver diary, June 15, 1936, Folder 498, Box 38, Series 401D, RG 1.1, RF Archives, RAC.

14. Dorothy Wrinch to Harry Sobotka, Sept. 17, 1937, Box 6, Folder 8.

15. Lord Todd and J. W. Cornforth, "Robert Robinson, 13 September 1886–8 February 1975," *Biographical Memoirs of Fellows of the Royal Society* 22 (Nov. 1976), 493.

16. Dorothy Wrinch to Warren Weaver, Sept. 24, 1937, Folder 499, Box 39, Series 401D, RG 1.1, RF Archives, RAC.

17. W. E. Tisdale, May 3, 1938, Folder 500, Box 39, Series 401D, RG 1.1, RF Archives, RAC.

18. W. E. Tisdale, Nov. 28, 1938, Folder 500, Box 39, Series 401D, RG 1.1, RF Archives, RAC.

19. Warren Weaver diary, Dec. 14, 1936, Folder 498, Box 38, Series 401D, RG 1.1, RF Archives, RAC.

CHAPTER 16

1. This scene and the next are drawn from Hager, *Force and Nature*; and from Warren Weaver's diaries, Linus Pauling's report on his meeting, and Weaver's confidential report to the Rockefeller Trustees, Folder 500, Box 39, Series 401D, RG 1.1, RF Archives, RAC.

2. Linus Pauling to Warren Weaver, February 23, 1938, ibid.

3. This scene is imagined, though the quoted words—all from Folder 500, Box 39, Series 401D, RG 1.1, RF Archives, RAC—are verbatim: WW to LP, April 11, 1938; WET to WW, April 25, 1938; LP to WW, February 23, 1938; WW to LP, March 2, 1938.

4. Pauling's remarks here and in the following paragraphs are taken from his report (dated March 31, 1938) on his meeting with Wrinch, and his cover letter to Weaver, dated April 11, 1938. The letter, the report, and Pauling's notes on his meeting with Wrinch are also in Folder 500, Box 39, Series 401D, RG 1.1, RF Archives, RAC.

5. In his written report he crossed "is" out and wrote in "seems to be."

6. "The Shape of Things Invisible"; "Trustees' Confidential Bulletin," April 1938, Folder 500, Box 39, Series 401D, RG 1.1, RF Archives, RAC.

7. I'm aware this will have to be shortened for the opera, but for now the long form will do.

8. Scene 3 is based on Warren Weaver's diaries and correspondence between Langmuir and Wrinch. I have invented Blogett's role in this scene, but I have not invented Blodgett.

9. Charles Tanford and Jacqueline Reynolds, *Nature's Robots: A History of Proteins* (Oxford University Press, 2001). Their account of the cyclol controversy is virulently anti-Wrinch.

10. Linus Pauling and Carl Niemann, "The Structure of Proteins," *Journal of the American Chemical Society* 61 (July 1939), 1860–1867.

11. Kay, *The Molecular Vision of Life*.

12. A. D. Booth to MS, private communication, March 2008.

13. Dorothy Wrinch to Eric Neville, n.d, DW Box 6, Folder 1.

14. Hager, *Force of Nature,* 230.

15. Ibid.

16. Michael Polanyi, *The Tacit Dimension* (University of Chicago Press, 1966).

17. Hager, *Force of Nature,* 240.

18. Lesk, "The Unreasonable Effectiveness of Mathematics in Molecular Biology."

19. Nat Hentoff, *A Doctor Among the Addicts* (Rand-McNally, 1968).

20. M. Senechal, "Narco Brat," in *Of Human Bondage,* ed. D. Patey, Smith College Studies in History, 52 (Smith College, 1998).

21. Warren Weaver, Tuesday, May 2, 1939, Folder 501, Box 39, Series 401D, RG 1.1, RF Archives, RAC.
22. Kay, *The Molecular Vision of Life*.

CHAPTER 17

1. Pieter Terpstra, *A Thousand and One Questions on Crystallographic Problems* (J. B. Wolters, 1952).
2. M. Senechal, "F Faces of Apatite and Its Morphology: Theory and Observation," *Journal of Crystal Growth* 78 (1986), 468–478 (with R. A. Terpstra, P. Bennema, P. Hartman, C. F. Woensdreght and W. G. Perdok).
3. Joseph Pitton de Tournefort, *The Origin and Formation of Stones* (1702).
4. R. J. Haüy, *Traite de Mineralogie*, 5 vols. (Delance, 1801).
5. Attributed to Aleksander Isaakovich Kitaigorodskii.
6. P. P. Ewald, "The Myth of Myths; Comments on P. Forman's Paper on 'The Discovery of the Diffraction of X-Rays in Crystals,'" *Archive for History of Exact Sciences* 6 (1969), 72–81.
7. For a detailed and definitive account of this experiment and all that led to it and from it, see A. Authier, *Early Days of X-Ray Crystallography* (Oxford University Press, 2012).
8. Biography of William Bragg, http://nobelprize.org/nobel_prizes/physics/laureates/1915/wh-bragg-bio.html.
9. Dame Kathleen Lonsdale, "Crystallography at the Royal Institution," in *Fifty Years of X-ray Diffraction*, ed. P. P. Ewald (Oesthoek, Utrecht, 1962), 412–413.
10. W. T. Astbury to J. D. Bernal, Sept. 12, 1928, JDB ADD 8287 J.2.
11. J. D. Bernal, handwritten note, JN B. 24.

CHAPTER 18

1. C. H. MacGillavry, *Symmetry Aspects of M. C. Escher's Periodic Drawings* (Oosthoek, 1965).
2. Alan Holden and Phyllis Singer, *Crystals and Crystal Growing* (Anchor Books, 1960; MIT Press, 1982).
3. M. M. Woolfson, *Direct Methods in Crystallography* (Oxford University Press, 1961).
4. Dorothy Crowfoot to J. D. Bernal, December 10, 1934 (or 1935?), JDB ADD 8287 J. 89.
5. Hans F. Jensen, *Insulin: Its Chemistry and Physiology* (Oxford University Press, 1938). Jensen studied with John Abel, the first to crystallize insulin. Jensen gives that date as 1922, though elsewhere 1926 is cited.
6. D. M. Wrinch, "On the Structure of Insulin," *Transactions of the Faraday Society* 33 (1937), 1368–1380.

7. Dorothy Crowfoot, "The Crystal Structure of Insulin. I. The Investigation of Air-Dried Insulin Crystals," *Proceedings of the Royal Society A* 164, 919 (Feb. 18, 1938), 580–602.

8. A. L. Patterson, "Experiences in Crystallography—1924 to Date," in *Fifty Years of X-Ray Diffraction*.

9. Linus Pauling to A. L. Patterson, February 15, 1939, LP.

10. Jack Morrell, *Science at Oxford, 1914–1939* (Clarendon Press, 1997).

11. WLB to J. D. Bernal, December 12, 1938, WLB 77M/68.

12. Warren Weaver, July 1, 1938, Folder 500, Box 39, Series 401D, RG 1.1, RF Archives, RAC.

13. W. L. Bragg, "Patterson Diagrams in Crystal Analysis," *Nature*, January 14, 1939, 73–74; J. D. Bernal, ibid., 74–75.

14. D. M. Wrinch, "The Geometry of Discrete Vector Maps," *Philosophical Magazine* 5.27, no. 180 (Jan. 1939), 98–122; D. M. Wrinch, "Vector Maps of Finite and Periodic Point Sets," *Philosophical Magazine* 5.27, no. 183 (April 1939), 490–507.

15. Dorothy Wrinch to Irving Langmuir, Oct. 4, 1938, DW Box 5, Folder 2.

16. Linus Pauling to David Harker, February 6, 1939, LP.

17. W. L. Bragg, "Patterson Diagrams in Crystal Analysis," *Nature*, January 14, 1939, 73–74.

18. H. Lipson and W. Cochran, *The Determination of Crystal Structures*, vol. 2, *The Crystalline State*, ed. W. L. Bragg (G. Bell and Sons, 1953).

19. I. Fankuchen to William V. Consolazio, National Science Foundation, October 7, 1958, IF.

20. J. Fridrichsons and A. Mathieson, "Image-seeking. A Brief Study of Its Scope and Comments on Certain Limitations," *Acta Crystallographica* 15 (1962), 1065.

21. B. M. Shchedrin, "Existence of a Solution of the Inverse Problem Estimating the Structure of Materials from Diffraction Data," *Computational Mathematics and Modeling* 12, no. 3 (2001), 243–251.

22. S. Skiena, W. Smith, and P. Lemke, "Reconstructing Sets from Interpoint Distances" (extended abstract, Symposium on Computational Geometry, 1990), 332–339; A. Lipson, S. G. Lipson, and H. Lipson, *Optical Physics*, 4th ed. (Cambridge University Press, 2011).

CHAPTER 19

1. D'Arcy Thompson's letter and the records of the fellowship are in the archives of Somerville College, Oxford.

2. Warren Weaver diary, March 14, 1938, Folder 500, Box 39, Series 401D, RG 1.1, RF Archives, RAC.

3. Lynda Grier to Dorothy Wrinch, October 6, 1938, Grier papers, LMH.

4. Warren Weaver diary, October 17, 1939, Folder 500, Box 39, Series 401D, RG 1.1, RF Archives, RAC.

5. Charles Singer to Dorothy Wrinch, June 1, 1954, DW Box 5, Folder 15.

6. D. Harker, "Colored Lattices," in *Structures of Matter and Patterns in Science*, ed. M. Senechal (Schenkman, 1980).

7. D. Harker and D. Wrinch, "Lengths and Strengths of Chemical Bonds," *Journal of Chemical Physics* 8 (1940), 502–503.

8. Dorothy Wrinch to Eric Neville, n.d. (fall 1940), DW Box 6, Folder 1. The letters she refers to are cited in LP: letter from LP to Harker July 6, 1940, and Harker's reply, July 16, 1940, both under LP Correspondence: Box #151.10, File: Harker, David.

9. Dorothy Wrinch to Eric Neville, March 29, 1941. All quotes in this paragraph are taken from Dorothy Wrinch's letters to Eric Neville, 1940 and 1941, DW Box 6, Folder 1.

10. Dorothy Wrinch, personal notes, n.d. (circa 1940), DW Box 6, Folder 6.

11. Dorothy Wrinch to Eric Neville, September 29, 1941, DW Box 6, Folder 1.

12. Margery Fry to Dorothy Wrinch, September 3, 1940, DW Box 4, Folder 3.

13. J. A. Fraser Roberts, "Reginald Ruggles Gates, 1882–1962," *Biographical Memoirs of Fellows of the Royal Society* 10 (Nov. 1964), 83–106.

14. Dorothy Wrinch to Eric Neville, December 10, 1940, DW Box 6, Folder 1.

15. See H. M. Miller log, Jan. 9, 1941, RF 1.1 401D B39 F502, RF Archives, RAC.

16. Karl Menger to Dorothy Wrinch, n.d., DW Box 7, Folder 3.

17. Otto Glaser to Dorothy Wrinch, December 14, 1940, DW Box 3, Folder 8.

18. Walter Stewart to Otto Glaser, January 6, 1941, ibid.

19. Dorothy Wrinch to Eric Neville, March 29, 1941, DW Box 6, Folder 1.

20. Dorothy Wrinch to Otto Glaser, May 3, 1941, DW Box 3, Folder 8.

21. Eric Neville to Dorothy Wrinch, May 22, 1941, DW Box 6, Folder 4.

22. F. B. Hanson, diary, August 1941, Folder 1565, Box 128, Series 200, RG 1.1, RF Archives, RAC.

23. Margery Fry to Dorothy Wrinch, September 23, 1941, DW Box 4, Folder 3.

24. Dorothy Wrinch to Eric Neville, September 29, 1941, DW Box 6, Folder 1.

25. D'Arcy Thompson to Dorothy Wrinch, March 18, 1943, DW Box 5, Folder 23.

26. Dorothy Hodgkin to Isidor Fankuchen, October 15, 1941, IF.

27. *New York Times*, August 21, 1941.

28. Dorothy Wrinch to President Herbert Davis, May 1942, Smith College Archives, Office of the President, Davis, Herbert John, 1940–1949, Box 10 of 10, Wilson College.

29. Dorothy Wrinch to Eric Neville, December 20, 1941, DW Box 6, Folder 1.

30. Otto Glaser to Dorothy Wrinch, August 2, 1942, DW Box 3, Folder 8.

31. Martin Buerger to Dorothy Wrinch, July 19, 1943, DW Box 4, Folder 14.

CHAPTER 20

1. For a full account of this sensitive episode in the history of Amherst College, see Douglas C. Wilson, "The Story in the Meiklejohn Files," in *Passages of Time: Narratives in the History of Amherst College*, ed. D. C. Wilson (Amherst College Press, 2007), 101–35.

2. Proceedings of the 50th annual meeting of the American Society of Zoologists, December 1953.

3. Grant to Amherst College for research in biology, Nov. 16, 1934, Folder 1563, Box 127, Series 200D, RG 1.1, RF Archives, RAC.

4. President Stanley King to the Amherst College Board of Trustees, May 29, 1944. Quoted by permission of the Trustees of Amherst College.

5. The programs, membership lists, and other materials from the Ladies of Amherst, 1912–1972, are held in the Special Collections of the Amherst College Archives and are open to the public. In 1972, the club renamed itself "The Women of Amherst." The college had just hired its first full-time female faculty member, and co-education was on the horizon. But a name change was not enough; times had changed too. The club folded soon afterward. The college appointed its first female president in 2011.

6. Amherst College Oral History Project Records, emeriti, Mrs. Theodore Soller, May 21, 1980, Box 2, Folder 29.

7. President Stanley King to the Amherst College Board of Trustees, May 29, 1944. Quoted by permission of the Trustees of Amherst College.

8. D. Fuess, *Stanley King of Amherst* (Columbia University Press, 1955).

9. Warren Weaver, Oral History, vol. 2, p. 258, interview 5, April 6, 1961, RG 13, RF Archives, RAC.

10. F. B. Hanson diary, Feb. 15, 1945, Folder 1566, Box 127, Series 200D, RG 1.1, RF Archives, RAC.

11. Folder 1565, Box 128, Series 200D, RG1.1, RF Archives, RAC.

12. F. B. Hanson diary, Nov. 14, 1944, Folder 1566, Box 127, Series 200D, RG 1.1, RF Archives, RAC.

13. Marshall Chadwell diary, October 24, 1947, RG 12, RF Archives, RAC.

14. Weaver did not mention the incident in his oral history interviews.

15. Smith College News Release, Friday, August 13, 1948.

16. G. Anslow and O. Glaser, "Copper and Ascidian Metamorphosis," *Journal of Experimental Zoology*, 111 (1949).

17. Carolyn Cohen, "Deciphering Protein Designs," in *Structures of Matter and Patterns in Science*, ed. M. Senechal, 77.

CHAPTER 21

1. Andrew Donald Booth to Dorothy Wrinch, June 29, 1946, DW Box 4, Folder 11.

2. Andrew Donald Booth to Warren Weaver, July 3, 1946, Folder 434, Box 34, Series 401D, RG 1.1, RF Archives, RAC.

3. Andrew Donald Booth to Dorothy Wrinch, August 12, 1946, DW Box 4, Folder 11.

4. W. N. Lipscomb, "Review of Dorothy Wrinch, Fourier Transforms and Structure Factors," *Journal of Physical and Colloid Chemistry* 51, no. 1215 (1947).

5. Berol, "Living Materials and the Structural Ideal."

6. J. von Neumann, "Letter to Norbert Weiner from John von Neumann," in *Proceedings of the Norbert Wiener Centenary Congress*, 1994, eds. V. Mandrekar and P. R. Masani (1997).

7. John von Neumann to Dorothy Wrinch, December 14, 1946, DW Box 7, Folder 5.

8. Dorothy Wrinch to Harlow Shapley, December 27, 1946, DW Box 7, Folder 5.

9. Dorothy Wrinch to Isidor Fankuchen, n.d. (1946), IF.

10. David Harker to the Guggenheim Foundation, 1947. Courtesy of the Guggenheim Foundation.

11. Irving Langmuir to the Guggenheim Foundation, 1947. Courtesy of the Guggenheim Foundation.

12. See LP.

13. Berol, "Living Materials and the Structural Ideal."

14. W. Cochran and M. Woolfson, "Have Hauptman & Karle Solved the Phase Problem?" *Acta Crystallographica* 7 (1954), 450–451.

15. Herbert Hauptman, interviewed by Istvan Hargittai, *Candid Science II: More Conversations with Famous Chemists*, ed. Magdolna Hargittai (Imperial College Press, 2003).

16. Senechal, *Structures of Matter and Patterns in Science.*

17. I received this and the next three reminiscences in reply to my query in Smith's alumnae e-newsletter, summer 2009.

18. I have found no evidence to support this rumor, but Russell's archivist says he has seen an inkling.

CHAPTER 22

1. A. Stoll, A. Hofmann, and T. Petrzilka, Die Konstitution der Mutterkornalkaloide. Struktur des Peptidteils. III. 24. Mitt. uber Mutterkornalkoloide. *Helvetica. chimica. Acta* 34, 1544 (1951).

2. R. Gordon Wasson, Albert Hofmann, and Carl A. P. Ruck, *The Road to Eleusis: Unveiling the Secret of the Mysteries*, 30th Anniversary ed. (North Atlantic Books), 2008.

3. M. Greene, *Natural Knowledge in Preclassical Antiquity* (Johns Hopkins University Press, 1992).

4. L. Růzička, "Arthur Stoll," *Biographical Memoirs of Fellows of the Royal Society*, (1972).

5. A. Stoll, "Recent Investigations on Ergot Alkaloids," *Progress in the Chemistry of Organic Natural Products*, 9th vol. (Springer-Verlag, 1952), 114–173.

6. In 1996, Sandoz merged with another Swiss mega-firm, Ciba-Geigy, to form Novartis, now one of Big Pharma's very biggest.

7. A. Hofmann, *LSD: My Problem Child* (McGraw-Hill, 1980).

8. For a more detailed account, see Lawrence Altman, *Who Goes First? The Story of Self-Experimentation in Medicine* (University of California Press, 1986).

9. Dorothy Wrinch, "'The Dark Ages' Scourge Led to Lifesaving Drugs," *New York Herald Tribune*, July 21, 1957. Reprinted in the Smith *Alumnae Quarterly*, February 1958, 82.

10. Dorothy Wrinch to Isidor Fankuchen, April 11, 1962, IF.

11. A. Wikler, *The Relation of Psychiatry to Pharmacology* (Williams and Wilkins, 1957).

12. Dorothy Wrinch, personal note, n.d., DW Box 29, Folder 2.

13. D. Wrinch, *Chemical Aspects of the Structures of Small Peptides: An Introduction* (Muksgaard, 1960).

14. Dorothy Wrinch to Harry Sobotka, n.d., DW Box 6, Folder 8.

15. Polanyi, *Personal Knowledge*, corrected edition (University of Chicago Press, 1962), 156.

16. Louis A. Cohen, book review, *Journal of the American Chemical Society* 83, no. 21, (1961), 4488.

17. H. N. Rydon, "The Cyclol Revived" (book review), *Nature*, July 8, 1961, 105.

18. Dorothy Wrinch, personal notes, n.d., DW Box 29, Folder 2.

19. Wrinch, *Chemical Aspects of Polypeptide Chain Structures and the Cyclol Theory* (Munksgaard, 1965).

20. See, e.g., H. S. Olcott, *Journal of the American Oil Chemists' Society*, 44 (Jan. 1967), 25A.

21. J. F. W. Herschel, *Familiar Lectures on Scientific Subjects* (Alexander Strahan, 1866), 412.

22. Heinrich Klüver, *Mescal and Mechanisms of Hallucinations* (University of Chicago Press, 1966), a reprint of both Klüver's 1928 book *Mescal: The "Divine" Plant and Its Psychological Effects* (Kegan Paul, Trench, Trubner) and his 1942 article "Mechanisms of Hallucination," first published in Q. McNemar and M. A. Merril, eds., *Studies in Personality* (McGraw-Hill), 175–207. In the preface to the reprints, Klüver explains that he published *Mescal* at the urging of C. K. Ogden, then editor of Kegan Paul's *Psyche* series. Evidently, Dot did not read it.

23. David Lewis-Williams, *The Mind in the Cave* (Thames and Hudson, 2002).

CHAPTER 23

1. P. N. Wrinch, The Military Strategy of Winston Churchill, *Studies in Political Science* 5 (Boston University, 1961).
2. H. Weyl, *Symmetry* (Princeton University Press, 1952).
3. H. Weyl, *Mind and Nature: Selected Writings on Philosophy, Mathematics, and Physics*, ed. P. Pesic (Princeton University Press, 2009).
4. Dorothy Wrinch to Warren Weaver, Sept. 24, 1937, Folder 499, Box 39, Series 401D, RG 1.1, RF Archives, RAC.
5. See, e.g., D. Wrinch, "The Twinning of Cryolite," *American Mineralogist* 37, nos. 3–4 (1952), 234–241; and "Some Observations on Twinning" (Symposium on Twinning, abstract 1, meeting of the American Crystallographic Association, June 1952).
6. Stress or pressure twins are formed after growth; they are another category altogether.
7. M. Senechal, "The Mechanism of Formation of Certain Growth Twins of the Penetration Type," *Neues Jahrb. Mineral. Monatsch* (1976), 518–525. Thirty-five years later, my ideas are still, as Dorothy would say, *sub judice*. "For this interesting theoretical model no experimental proof is yet available, but it appears rather reasonable," say Th. Hahn and H. Klapper, in "Twinning of Crystals," *International Tables for Crystallography*, Volume D, *Physical Properties of Crystals*, 2003. See also my recent suggestion, "Crystals Twins Revisited," *Israel Journal of Chemistry*, Special Issue: Quasicrystals, 51, nos. 11–12 (Dec. 2011), 1153–1155.
8. M. Senechal, "From Symmetry to Disorder: A Personal Odyssey," the 1991 Katharine Asher Engel Lecture, *Five College Ink*, Five Colleges, Inc., Amherst, MA, May 1991.
9. Senechal, ed., *Structures of Matter and Patterns in Science*.
10. Italo Calvino, *Invisible Cities*, trans. William Weaver (Harcourt Brace Jovanovich, 1972), 86.
11. Lipson, *Optical Physics*.
12. Denis Noble, foreword to *The Philosophical Foundations of Neuroscience*, ed. M. R. Bennett and P. M. S. Hacker (Blackwell, 2003).

CHAPTER 24

1. Shechtman, D., I. Blech, I. D. Gratias, and J. W. Cahn, "Metallic Phase with Long-Range Orientational Order and No Translational Symmetry," *Phys. Rev. Lett.*, 53, 1951 (1984).
2. W. B. Yeats, "Among School Children," in *The Tower* (Macmillan, 1928).
3. "Dan Shechtman and a Revolution in Basic Science," Technion website, April 1, 2011, http://technionlive.blogspot.com/2011/04/serious-matter-dan-shechtman-and.html.

4. "The Nobel Prize in Chemistry 2011—Presentation Speech," Nobelprize. org., Dec. 23, 2011, http://www.nobelprize.org/nobel_prizes/chemistry/laureates/2011/presentation-speech.html.

5. Harriet Ward, *A Man of Small Importance: My Father, Griffin Barry* (Dormouse Books, 2003).

6. Dorothy Wrinch to Dora Russell, October 14, 1965, Dora Russell Papers, Internationaal Instituut voor Sociale Geschiedneis, Amsterdam.

APPENDIX

1. Hadamard seems unaware that this paper, published in 1923, was one part of DW's prizewinning 1918 Girton thesis on transfinite types.—MS

INDEX

Note: DW is, of course, Dorothy Wrinch. I break with indexing convention by alphabetizing DW as D.